Palgrave Macmillan Transnational History Series

Series Editors
Akira Iriye
Harvard University
Cambridge, USA

Rana Mitter
Department of History
University of Oxford
Oxford, UK

This distinguished series seeks to develop scholarship on the transnational connections of societies and peoples in the nineteenth and twentieth centuries; provide a forum in which work on transnational history from different periods, subjects, and regions of the world can be brought together in fruitful connection; and explore the theoretical and methodological links between transnational and other related approaches such as comparative history and world history.

More information about this series at
http://www.palgrave.com/gp/series/14675

Ignacio Siles

A Transnational History of the Internet in Central America, 1985–2000

Networks, Integration, and Development

Ignacio Siles
University of Costa Rica
San Jose, Costa Rica

Palgrave Macmillan Transnational History Series
ISBN 978-3-030-48946-5   ISBN 978-3-030-48947-2   (eBook)
https://doi.org/10.1007/978-3-030-48947-2

Cover illustration: Galina Tolochko / Alamy Stock Photo

This Palgrave Macmillan imprint is published by the registered company Springer Nature Switzerland AG.
The registered company address is: Gewerbestrasse 11, 6330 Cham, Switzerland

For Lea

ACKNOWLEDGMENTS

Book writing is always an exercise in constructing networks. Gathering data, carrying out interviews, finding additional sources of information, and producing a manuscript all require weaving together an infinite number of ties. It has been a pleasure to work with everyone at Palgrave Macmillan. I am grateful to Megan Laddusaw, the editors of the Palgrave Macmillan Transnational History Series, and anonymous reviewers for their help in improving this manuscript.

An indispensable node for carrying out this project has been the Research Center in Communication (CICOM) at Universidad de Costa Rica (UCR), where I received the needed logistic and administrative support, as well as the chance to discuss the research's progress with colleagues. I am especially grateful to Patricia Vega and Yanet Martínez, CICOM's directors during the duration of this project, for having created the proper conditions to work on it. I also extend my thanks to the School of Communication at UCR, notably to its director, José Luis Arce, for making evident his commitment to building bridges between teaching and research. Fernando García, Research Vice-President at UCR, offered me his invaluable support to conduct this project. I could not have published this book so quickly without his help. I am grateful to Sonya Kozicki-Jones and Kelsey Peterson for their excellent work in the translation of this manuscript.

A number of organizations and individuals contributed to constructing the network that is this book. I am indebted to each person interviewed over the last fifteen years for their willingness to share their stories with me.

I hope that in this book they find a faithful reflection of their experiences. I am grateful to Abel Brenes, Ana Lucía Chavarría, Guy de Téramond, Mario Guerra, and Allan Ruiz for their permission to use various images as part of this book. Recognition in particular is due to Rachel Plotnick, Fabián Prieto and Carlos Sandoval, all of whom gave me their views on preliminary drafts of the manuscript. Niels Brügger and Valérie Schafer, as editors of *Internet Histories*, also helped to strengthen certain claims. Mariana Álvarez, Alejandra Mora and Susan Leitón served as assistants at various moments of the project; my thanks to them both for their kindness, availability, and enthusiasm while collaborating with me. Virginia Mora repeatedly invited me to her Media History course in order to share findings related to this project. Some preliminary results were also presented at the Association of Internet Researchers (Montreal, 2018) and at the Association of Latin American Communication Researchers (San José, 2018), where I received opportune comments for improving this work.

I likewise wish to recognize the continuous support of NIC Internet Costa Rica, notably that of María Antonieta Chavarría, to make contact with many people throughout the Central American region. UCR's library system (SIBDI) was invaluable for obtaining bibliographical materials, as were CIDACS staff members at UCR's Instituto de Investigaciones Sociales and INTAL-LIB, who helped me obtain files that were hard to find online. René Medina, at RACSA, provided valuable help in making contact with key sources.

This book is dedicated to my daughter Lea, the border-crossing transnational: born in Chicago, she inherited Costa Rican, Nicaraguan, Guatemalan, and Uruguayan roots from her parents and grandparents. I am grateful to Lea for her love. Finally, as always, I thank Tania for allowing me to accompany her on this voyage.

CONTENTS

ABOUT THE AUTHOR

Ignacio Siles is a professor of media and technology studies in the School of Communication at the University of Costa Rica. He is a researcher in the Centro de Investigación en Comunicación (CICOM) at the same institution. He holds a PhD in Media, Technology and Society (Northwestern University, USA) and a master's degree in Communication (Université de Montréal, Canada). He is the author of *Networked Selves: Trajectories of Blogging in the United States and France* (Peter Lang, 2017) and *Por un sueño en.red.ado: Una historia de Internet en Costa Rica* (EUCR, 2008) along with several articles on the relationship between technology, communication, and society.

Acronyms

ACDI	Canadian International Development Agency
ANTEL	The National Telecommunications Administration (El Salvador)
APC	Association for Progressive Communications
BCIE	Central American Bank for Economic Integration
BID	Inter-American Development Bank
CATIE	Tropical Agricultural Research and Higher Education Center
CCITT	International Telegraph and Telephone Consultative Committee (ITTCC)
CERN	European Organization for Nuclear Research
COHCIT	National Council for Science and Technology (Honduras)
COMTELCA	Regional Technical Telecommunications Commission (Honduras)
CONACYT	National Council for Science and Technology (El Salvador)
CONCYT	National Council for Science and Technology (Guatemala)
CONICIT	National Council for Scientific and Technological Investigation (Costa Rica)
CRIES	Regional Coordinator of Economic and Social Investigation, Nicaragua NGO Nongovernmental organization
CRNet	Costa Rican Research Network/National Investigation Network (Costa Rica)
CSUCA	Superior Central American University Council
EARN	European Academic Research Network
ECLAC	Economic Commission for Latin America and the Caribbean
ENITEL	Nicaraguan Telecommunications Enterprise
GUATEL	Guatemalan Telecommunications Enterprise

HONDUTEL	Honduran Telecommunications Enterprise
IANA	Internet Assigned Numbers Authority
ICE	Costa Rican Electricity Institute
ICM	International Connections Management
ICT	Costa Rican Tourism Institute
ICT4D	Information and Communications Technologies for Development
IDA	International Development Agency
IDRC	International Development Research Centre (Canada)
IGC	Institute for Global Communications
INTEL	National Telecommunications Institute in Panama
Intelsat	International Telecommunications Satellite Organization
ITCR	Costa Rica Institute of Technology
ITU	International Telecommunication Union
MCC	Central American Common Market
NSRC	Network Startup Research Center
PanAmSat	Pan American Satellite corporation
RACSA	Costa Rican Radiographic Institution
RAIN	Nicaraguan Academic Internet Network
REDCSUCA	Central American University Network for Scientific Information
RedHUCyT	Hemisphere-Wide Inter-University Scientific and Technological Information Network
SENACYT	National Secretary of Science, Technology, and Innovation
SICA	Central American Integration System
SNDP	Sustainable Development Networking Programme
TELCOR	Nicaraguan Institute of Telecommunications and Postal Services
TELGUA	Guatemala Telecommunications
UCA	José Simeón Cañas Central American University (El Salvador)
UES	University of El Salvador
UNAH	National Autonomous University of Honduras
UNDP	United Nations Development Programme
UNED	National Distance Education University (Costa Rica)
UNI	National University of Engineering (Nicaragua)
USAID	United States Agency for International Development
UTP	Technological University of Panama
UVG	Guatemala's University of the Valley (of Guatemala)

LIST OF FIGURES

LIST OF TABLES

CHAPTER 1

"Follow the Networks"

Abstract This chapter sets the stage by introducing the historical relevance and intellectual significance of the history of the Internet in Central
America. It unpacks the theoretical and methodological underpinnings of
the book. To this end, the chapter introduces three important issues. First,
it argues for examining the history of the Internet through a transnational
lens (and clarifies what this means for the project). Second, it theorizes
technology (such as computer networks) as a political project of regional
integration. Third, it puts forth a view of technology and development as
mutually constitutive. Methodologically, the analysis draws on archival
research and 80 interviews with protagonists of networking initiatives.
The chapter concludes by offering an overview of the book.

Keywords Central America • Computer networks • Development •
Internet history • Integration • Transnational history

On February 27, 1994, three Costa Rican engineers took an afternoon
flight from San José to Managua, Nicaragua. The timing for this trip was
good in more ways than one. Little by little, more than a decade of war in
the region was coming to an end. The trip had a single purpose: participate in Nicaragua's connection to the Internet. In Managua, a group of
collaborators who had worked for months to establish this link awaited
them. For almost three years, they had been making plans together for

© The Author(s) 2020   1
I. Siles, *A Transnational History of the Internet in Central America,*
1985–2000, Palgrave Macmillan Transnational History Series,
https://doi.org/10.1007/978-3-030-48947-2_1

Nicaragua's Internet connection through Costa Rica via an analog microwave link built in the late 60s, a decade in which the concept of Central American integration had flourished. From Costa Rica, Nicaragua would be connected to Homestead, Florida through a satellite antenna. This goal was achieved the very next day and was celebrated enthusiastically. A public event was held at the Nicaraguan university that led this initiative. After a series of training and work sessions with their Nicaraguan counterparts, the Costa Rican engineers returned to San José on March 2. Only four months later, they would repeat this process in a different setting: the new site was Panama, but the purpose and procedures were almost identical.

This story has captivated me since I first heard it a few years ago for various reasons. First, because of its historical importance. Between 1993 and 1996, Central American countries established direct links to the Internet for the first time in history. Moreover, the Internet connection of one country through the infrastructure of another was a technological milestone in Latin America. Second, it reflects the creativity of a group of people from countries with few economic resources, in a region devastated by years of war and crisis. Peace agreements negotiated at the end of the 1980s required lengthy processes of economic, political, and social reconstruction over the following decade (Pérez Brignoli, 2010). In most countries of the region, peace was only attained well into the 1990s. Thus, because of the historical context in which this occurred, connecting to the Internet was a political as much as a technological achievement. Third, the significance of the story also involves the actors in question: the network linked two neighboring countries with a complicated historical relationship marked by chronic controversies. Finally, this achievement matters for how it symbolizes a specific era in the history of computer networks in the southern hemisphere. Various actors in Latin America experienced similar processes in some way or another. Establishing new nodes of computer networks required these kinds of exchanges and flows, these twists and turns.

How and why did Central America connect to the Internet? What consequences did the link to early computer networks have in the region? This book sets out to answer these two research questions through an original analysis of the projects that resulted in the first Internet connections in countries of the region. These projects were characterized by the establishment of not only technological networks but also transnational collaborations between actors and organizations. In this way, Central American countries connected *to* and *through* computer networks such as the Internet. Drawing on archival work and interviews with the protagonists of these projects (including directors and collaborators of the networking projects, figures in politics, government and international organizations, representatives of telecommunications companies, and pioneer users, among others), this book examines how initiatives to connect to early computer networks unfolded and were developed from the mid-1980s to the end of the 1990s.

The following pages discuss the early development of the Internet in a region that has not received much academic attention. Consistent with the tendency to provide "hagiographic" descriptions of successful cases (Russell, 2017), historical research has primarily dealt with the most connected countries. As a result, we know little of how the Internet has been historically envisioned and implemented in less connected regions, such as Central America. Therefore, our understanding of the early development of computer networks in the global south is limited. Over the past few years, there has been a growing interest in studying the use and development of media technologies in Latin America (Chan, 2014; Kleine, 2013; Medina, 2011; Penix-Tadsen, 2016; Takhteyev, 2012). Together, these monographs seek to "move the story of invention and innovation southward; study forms of local innovation and use; analyze the circulation of ideas, people, and artifacts in local and global networks; and investigate the creation of hybrid technologies and forms of knowledge production" (Medina, da Costa Marques, & Holmes, 2014, p. 3). However, these studies tend to focus on some of the largest countries in the southern continent. This book suggests that, because of its history and political, economic, and social configurations, the study of Central America can also offer important analytic lessons for interdisciplinary research on the development of media technologies, including research conducted in and about Latin America.

The possibility of establishing communication networks through technology has raised hopes throughout history. Mattelart (2000) traces this process back to at least the eighteenth century, when the network became "the emblematic figure of the new organization of society" (p. 15). Few concepts have marked the turn of the century more than the "network" (Boltanski & Chiapello, 1999; Castells, 1996). In a recent interview, Guy de Téramond, one of the architects and protagonists of the story of how Costa Rica and Nicaragua came to be interconnected, which this book starts with, offered a terse but profound explanation of the motivations that characterized these types of projects: "Such is the nature of networks"

(Siles, 2017a, p. 352). There are several ways to interpret this assertion. Seen strictly as a computational phenomenon, de Téramond suggests that networks require new nodes in order to acquire or enhance their value. A more radical vision would attribute networks with a natural potential for expansion. Taken together, these interpretations capture a common understanding of the capacity that was ascribed to the Internet at the dawn of Central America's interconnection: networks have an intrinsic capacity to enhance integration and collaboration. In other words, integration would be a natural result of the construction of networks. This book transforms this assumption into an empirical question.

To that end, I propose to adapt actor-network theory's classic tenet ("follow the actors"). Actor-network theory considers every "fact" as a network composed of human and non-human actors that assume identities through a multiplicity of negotiations and interaction strategies. "Following the actors" thus means "[catching] up with [actors'] often wild innovations in order to learn from them what collective existence has become in their hands [...] which accounts could best define the new associations that they have been forced to establish" (Latour, 2005, p. 12). Over the following pages, I show that the study of computer networks also requires "following the networks." First, this means tracking the processes through which computer networks arrived in Central America in the mid-1980s. Second, this tenet invites us to understand how different nodes emerged in different parts of the Central American region; how flows of exchanges between these nodes were established; and through which actors, logic, and contexts these exchanges became possible. This is, in essence, an exercise in transnational analysis. That is the task set out for this book.

Comparative work about Costa Rica, El Salvador, Guatemala, Honduras, and Nicaragua has been frequent in academic research. However, deciding what counts as part of the "Central American region" is an exercise fraught with tensions. In this project, Panama was included in the analysis in addition to the five other countries, due to both its participation in the processes analyzed as well as to its historical connections with the processes described in the present work. For that reason, whenever Central America is mentioned, I refer to the aforementioned nations (i.e. *América Central*), despite the clear links of the study with the strictly delimited geographic region's history (i.e. *Centroamérica*). In contrast, Belize was excluded from the investigation. Although it could be argued that geographically it is part of this region, and that there was some collaboration between the actors discussed in this book and their counterparts in Belize, their networking processes were somewhat different compared to the other cases examined here.

NETWORKS, INTEGRATION AND DEVELOPMENT: THEORETICAL AND METHODOLOGICAL CONSIDERATIONS

By examining the early process of the connection to computer networks in Central America, this book dialogues with different interdisciplinary fields of knowledge in order to develop three theoretical arguments. These arguments span technology research, integration processes, and perspectives on development.

A Transnational History of the Internet

First, I argue that transnational flows of knowledge, data, and technologies are not only an inherent feature of the Internet, but rather a constitutive characteristic of its historical development. This is crucial to understand the histories of the Internet, but it has been seldom recognized in scholarly literature. Most historical research has focused on the study of the Internet mainly through national accounts (Brügger & Milligan, 2018; Goggin & McLelland, 2017). Transnational, regional networking efforts have received significantly less academic attention. This book makes visible the importance of transnational processes in the history of the Internet.

The study of transnational histories gained traction at the turn of the century, in the context of marked academic concerns about processes such as globalization. Transnational history is more an "umbrella" term than a field with established conceptual boundaries. As such, it tends to be defined more as an approach, "a way of seeing" things (Beckert, cited in Bayly et al., 2006, p. 1454), "an angle, a perspective" (Iriye & Saunier, 2009, p. xx), rather than as a specific method or theory. The stream of studies encompassed by the notion of transnational history can be discussed in terms of three basic orientations.

First, transnational histories emphasize the study of certain *objects* or, more precisely, certain processes, namely, flows, circulation, movements, connections, and exchanges that "operate over, across, through, beyond, above, under, or in-between polities and societies" (Iriye & Saunier, 2009, p. xviii). What passes through and crosses borders are people, knowledge, technologies, ideas, practices, and institutions. Thus, Hofmeyr argues that "the key claim of any transnational approach is its central concern with movements, flows, and circulation, not simply as a theme or motif but as an analytic set of methods which defines the endeavor itself" (cited in Bayly et al., 2006, p. 1444). Rather than abandoning the focus on political or geographical constructs such as countries, this approach complements it through a study of common processes that connect them and, thereby, redefine them. In this particular way, transnational history intersects with cultural studies and their interest in mechanisms of circulation.

Second, transnational histories highlight the work of specific *actors* and how they obtain transnational agency. Thus, studies have focused on the work of international organizations (Iriye, 2002; Keohane & Nye, 1972). Van der Vleuten (2006) contends that the analysis of these actors allows us to understand "how transnational networks were built, how divisions of labor between international organizations, state agencies and private companies were negotiated, and even how transnational linking processes failed" (p. 305). These organizations are analyzed as sites of internal tension (van der Vleuten, Anastasiadou, Lagendijk, & Schipper, 2007). Clavin (2005) also maintains that the role of these organizations should be understood in light of their interaction with government operations. To explain the agency of these specific kinds of actors, studies have highlighted some of their main organizational and coordination dynamics, most notably how these shape networks. Transnational histories are thus characterized by the study of networks made up of a variety of actors (Kohlrausch & Trischler, 2014; Snyder, 2011). Researchers have also stressed the role of so-called "network entrepreneurs," individuals who work to bring together previously separated actors and groups (Burt, 2000). The premise that underlies this approach is that the transnational circulation and flow of people, ideas, or products requires flexible structures that endow this flow with vitality and reach.

Third, scholars have highlighted the specific *consequences* that can be associated with the study of transnational processes. Empirically, one of the main contributions of this group of studies has been to demonstrate the importance of these flows in the historical formation of the nation-state or regionalism processes. For Bayly, "the 'nations' embedded in the term 'transnational' were not originative elements to be 'transcended' [...] Rather, they were the products—and often rather late products—of those very processes" (cited in Bayly et al., 2006, p. 1449). For this reason, transnational approaches have been key in recent analyses of Europe's historical construction (Kohlrausch & Trischler, 2014; van der Vleuten & Kaijser, 2006). Conceptually, scholars argue that a significant consequence of transnational histories is a renewed understanding of pre-established notions or, as Connelly puts it, a desire to "challenge ossified categories" (cited in Bayly et al., 2006, p. 1447). These categories tend to be replaced by notions that emphasize the "lived history" that infuses them (van der Vleuten, 2008, p. 984). More generally, what this approach allows then is "a new and more accurate perspective on existing themes in historical scholarship, a novel understanding of not only global or regional integration issues but also national and local history" (van der Vleuten, 2008, p. 987).

In the case analyzed in this book, a transnational approach makes visible the processes of network formation through which people, knowledge, and technologies circulated among Central American countries, and the role of specific actors (such as international organizations and "network entrepreneurs") through which the Internet became a reality in the region during the 1990s. This book argues that this perspective is indispensable for rethinking the history (or histories) of the Internet as a technology.

By emphasizing the formation of exchange networks, I do not mean to suggest that these collaborations were devoid of controversy. Ambiguities, tensions, and conflicts are an intrinsic part of the establishment of networks and technological projects. This book thus recognizes and examines the differences which emerged between countries and organizations, as well as the disputes that emerged within each country in Central America that pursued networking initiatives. Clavin (2005) reminds us that, once established, networks are not perpetual; on the contrary, they can be replaced at any time. I also avoid hailing the establishment of transnational networks as an intrinsically positive process. Mattelart (2000) contends that "networks have never ceased to be at the center of struggles for control of the world" (p. viii). The cases of "transnational states" and "transnational elites" demonstrate that the establishment of these types of networks has also limited, rather than enabled, the development of regions such as Central America (Bull, 2005; Sánchez-Ancochea & Martí i Puig, 2014).

Technology as a Political Integration Project

A second argument of this book is that as the Central American political and economic crisis of the 1980s came to an end, establishing computer networks brought about political projects of regional integration. In inter-disciplinary fields such Science and Technology Studies (STS), several authors have shown that technology is not neutral, but rather materializes a political project in itself. In a classic essay, Winner (1980) pointed out that technology is political not only because it is designed for specific pur-poses, but also because its conception seeks to make certain forms of social organization imperative. Technology thus creates a social system that legitimizes certain world views (expressed through specific uses of that technology), while sanctioning others. The use of technology thus becomes a terrain of tensions. Users can accept the values inscribed in technology, but they can also "interpret, challenge, reject, and modify" the political script they contain (Gillespie, 2007, p. 89). This perspective has found fertile ground in contemporary studies that examine media technologies as more than mere carriers of symbolic messages.

The premise behind this body of work is that technology and society are mutually constitutive and, therefore, are inseparable. Another way of expressing this idea is through the notions of "socio-technical" and "het-erogeneous" engineering (Law, 1987; Law & Callon, 1988; Morris, 2009). These notions suggest that the development and establishment of technology is a process in which a variety of heterogeneous elements are woven through the formation of "networks," "seamless webs" or "sys-tems" that are as technical as they are social (Hughes, 1983; Latour, 1991). Hughes (1986) summarizes this position: "The technological sys-tems of the system builders, such as an electric-light and power system, interconnect components so diverse as physical artefacts, mines, manufac-turing firms, utility companies, academic research and development labo-ratories, and investment banks" (p. 287).

In a similar manner, although it seems like a purely "technical" issue, implementing a technology such as the Internet (like projects examined in this book did) required the articulation of a heterogeneous network of elements, which included knowledge about matters related to the design and operation of computer protocols, but also the search for funds, nego-tiating with governments, academic authorities and state entities, inter-preting telecommunication monopoly laws and regulations, building alliances with international organizations, recruiting and training collabo-rators, and working with counterparts abroad to enhance the reach of the network, among other issues.

To make sense of these processes, I draw upon the work of Eden Medina (2011) about the relationship between technology and politics in

Allende's Chile: "technology can help scholars understand historical and political processes" (p. 8). In the case of this study, one of these processes related to Central American integration. Integration is a multidimensional concept. It is both a project and its products, a context that provides logic, discourse, and orientations to processes. Integration is usually understood and studied in economic or political terms. At the economic level, integration is interpreted as a matter of intra-regional trade in the context of international markets and structural factors. In the political sphere, scholars typically stress relations of power and hegemony between governments: "integration arises primarily as a result of the convergence of preferences in the larger states of a region [...] [which] co-operate and get involved in asymmetrical bargaining" (Sánchez Sánchez, 2009, p. 5). Integration is usually understood as a continuum of positions ranging from the definition of common goals, reforms, and agendas aimed at favoring some degree of convergence of preferences for the collective benefit—at one end—to the formation of a single political or economic entity—at the other end (CEPAL-BID, 1997).

Further to this, the present work develops an approach to integration that emphasizes the notion of interconnection. I argue that technology (as a political project) can materialize and enact specific notions of integration. Thus, both the implementation of computer networks and integration processes can be seen as products of heterogeneous engineering. The actors studied in this book envisioned integration as the formation of sociotechnical networks that could enable the transnational circulation of people, knowledge, and technologies towards common goal: the connection of Central American countries to computer networks as a means to foster development.

If "technologies are historical texts," as Medina points out (2011, p. 8), then the analysis of networking processes allows for a better understanding of a relatively unknown part of Central American history. This book offers an original and inedited account of a key decade in the history of Central America seen through the lens of technology. It is difficult to find mention of technological projects in the literature on Central American integration, let alone analytical considerations of their significance. This book contributes to this body of work by arguing that technology is a crucial way of critically examining the notion of integration and the visions of Central America that were associated with it at the end of the 1980s crisis.

A similar project has been conducted over more than a decade in Europe. These studies posit a parallel development of technological systems and various definitions of Europe and its integration (Kaijser, van der Vleuten, & Högselius, 2016; Schot & Scranton, 2014; van der Vleuten & Kaijser, 2006). The main lesson that can be drawn from this body of research is that the definitions of technology (such as the Internet) and regions (such as Central America) should not be taken for granted but rather be seen as mutually constitutive. In this book, I argue that, while political and economic integration efforts since the 1960s made possible the networking initiatives of the 1990s, these initiatives also helped materialize or express an approach to Central American integration that has seldom been examined and recognized. In an era in which the need to protect borders has often been defended, remembering a historical moment that sought to overcome them is also, in essence, a political act in itself.

Sociotechnical Configurations of Development

A long-standing premise in the history of technological networks is that interconnection through communication infrastructures leads to development (Mattelart, 2000). The most significant integration efforts in Central America of the twentieth century (in particular the 1960s and 1990s) were devised to achieve that specific goal. International organizations that contributed funds to networking projects in the region did so with this objective in mind as well. The third theoretical argument of this book connects the analysis of networking projects with efforts to understand their implications for regional development. The analysis of the link between computer networks and development can be explored at the institutional level, in the sense of sociology's "new institutionalism." This perspective privileges the study of institutions, which are defined as those "regulative, normative, and cultural-cognitive elements that, together with associated activities and resources, provide stability and meaning to social life" (Scott, 2014, p. 56). In this way, I provide an alternative to studies that seek answers and explanations at the level of access divides and that, in this way, "isolat[e] [technologies] from their much broader economic and social context" (Hoffmann, 2004, p. 1).

The link between technology and development connects the discussion of the history of computer networks in Central America with a variety of disciplinary fields. Studies in Information and Communication

Technologies for Development (ICT4D), for example, have provided numerous contributions to theorize this link from an institutional or systemic perspective (Bass, Nicholson, & Subrahmanian, 2013; Hoffmann, 2004; Kleine, 2013; Silva & Figueroa, 2002). They have sought to transcend the discussion of development based on measures of economic growth, to focus instead on people-centered processes and the possibilities they have to "lead the lives they have reason to value" (Sen, 1999, p. 3). However, the studies associated with this tradition have been limited in two main ways. First, they tend to conduct research into specific technology implementation initiatives rather than the general dynamics that can enable (or limit) development through technology. Second, these initiatives are studied mainly at the country level, and thus development at the regional level (entailed by processes such as integration) is usually not considered.

This book situates concrete networking initiatives within the context of a field, ecology, or wider sector, which makes it possible to understand how favorable conditions are created to achieve longer-term development at national and regional levels. To this end, I propose to analyze the historical development of computer networks such as the Internet through three related dimensions or levels:

- *Sociotechnical systems and networks*: This level privileges the study of computer networks as cultural artifacts. As noted above, the development of these artifacts requires articulating different elements through sociotechnical networks that give support and political value to systems and infrastructures. As this process takes shape, these systems and infrastructures become essential for actors. The articulation of these networks also involves putting into action specific and mutually constitutive views of technology and society (Gillespie, 2007; Latour, 2005; Law, 1987).
- *Institutional arrangements*: This level emphasizes the study of how institutional fields such as telecommunications are formed, maintained, diffused, and debilitated. These fields are comprised of organizations, norms, laws, knowledge, politics, and forces, as well as types of discourse and logic. This level allows computer networks to be envisioned as part of organizational, legal, and regulatory frameworks that, through the establishment of certain arrangements and provisions, can be difficult to modify (Hoffmann, 2004; Mansell, 2001; Silva & Figueroa, 2002).

– *Sociocultural configurations.* This level stresses the study of computer networks as cultures in themselves where contents, symbolic texts, appropriation and creation practices circulate. Users of these technologies are seen as having historically and culturally situated agency. These practices and contents lead to the emergence of "media cultures" or "digital cultures." Privileging the study of computer networks as media technologies also invites an assessment of how meanings and notions emerge around them that become fundamental parts of culture (Couldry, 2012; Hepp, 2012; Sassen, 2017; Siles, 2017b).

This book examines the trajectory of computer networks in Central America by considering them simultaneously as sociotechnical systems, institutional arrangements, and sociocultural configurations. Thus, it does not focus on specific technologies, public policies, or the uses of these technologies as isolated phenomena. Instead, it analyzes how these three dimensions converge in specific conceptions or views of development, while also examining development issues from the field and sphere of influence that have formed around technologies such as the Internet.

Another contribution of this book is to incorporate the formation of theories as a vital part of this institutional field. Thus, I provide an archeology—in the sense given by Michel Foucault, that is, an analysis of the conditions that have made it possible for certain discursive elements and forms of knowledge to emerge (Davidson, 1986)—of one of the first theories devoted to the implications of technology for development, which took shape in Latin American at the turn of the century.

Finally, I situate the process of the privatization of telecommunications in the 1990s as part of the realm in which computer networks took off in Central America. I thus explore the institutional tensions that privatization processes brought about: on the one hand, they created conditions for the formation of an incipient industry in some countries, and facilitated the emergence of early "digital cultures"; on the other, they hampered the pursuit of the integrationist dream, and limited the scope and potential of computer networks for regional development. I argue that the implications of networking projects to foster the development of Central America must be understood in the context of this tension.

A Note on Method

The analysis presented in this book comes mainly from two sources of data. First, archival research with primary sources was carried out over several years. Numerous documents related to the historical implementation of the Internet were collected in Central American countries. Most of these documents were preserved by the protagonists of networking initiatives or in the archives of organizations such as NIC Internet Costa Rica. It is important to point out that, despite the fragmentation involved in studying a process ranging over six countries, the digitization of some of these documents made tracking, consulting, and analyzing these files possible.

Second, a total of 80 interviews were conducted. These interviews were carried out in two different stages. I held an initial group of 44 interviews between 2005 and 2006. These interviews focused primarily on the Costa Rican case. Costa Rica was the first Central American country to connect to the Internet, and it also played an important role in the expansion of computer networks in the region (Siles, 2008). These conversations were thus intended to obtain a clearer picture of the local networking process in this country. Building on that first stage of data collection, a second round of 36 interviews was conducted between 2017 and 2020. This stage concentrated mostly on networking projects in other Central American countries. It sought to better understand the dynamics through which networks such as X.25, BITNET, and the Internet acquired regional nodes and users.

These interviews reflect the systemic or institutional approach described previously. In other words, I interviewed a multiplicity of actors with different roles in networking processes. I talked to engineers, coordinators, and collaborators of these projects, but also to promoters and early users, representatives of state telecommunications companies, actors from international, state and non-governmental organizations, and to some of the academics who first became interested in the Internet as an object of study in the 90s.

THE NODES OF THIS JOURNEY

Before delving deeper into the evidence that supports the arguments developed in the book, it is necessary to more carefully consider the context that made it possible for Central American countries to connect to computer networks. To this end, Chap. 2 discusses political, economic, and social processes that took place in the region in the second half of the twentieth century. It examines regional integration initiatives developed in the 1960s (i.e., the creation of the Central American Common Market) and in the 1990s (i.e., the Central American Integration System). This chapter argues for understanding the mutual configuration between regional integration processes and technological projects (such as roads, microwave analog links, and computer networks).

Chapter 3 begins the empirical analysis of the networking experiences and efforts that preceded access to the Internet in the region. The notion of "founding networks" allows for discussion of two important processes. On the one hand, the chapter analyzes projects for connecting to early computer networks in the region. These projects promoted political visions that became reality by means of the use of different technologies (i.e., X.25, UUCP, and BITNET). On the other hand, the term "founding" is used in the chapter to discuss the formation of transnational networks of collaborative efforts between people in a number of the region's countries. In some cases, these networks had a longer-lasting effect in the region than did early computer networks.

Once these "founding networks" were implemented, the next goal was to actually connect to the Internet. This required enabling technological access points in the Central American region. Chapter 4 examines the "regimes of alliances" (Gillespie, 2007) between a variety of organizations that were formed to make this happen. These organizations included, for example, the National Science Foundation (NSF), which sought to facilitate the use of the Internet outside the United States. The emergence of PanAmSat constituted a crucial development in the formation of this "regime of alliances." Actors such as the Organization of American States (OAS) and the United Nations Development Program (UNDP) mobilized their political leverage to promote the connection to computer networks in the region.

Chapter 5 analyzes in detail how each country of the region connected to the Internet: Costa Rica (1993), Nicaragua (1994), Panama (1994), Honduras (1995), Guatemala (1995), and El Salvador (1996). The discussion follows two parallel processes. On the one hand, the chapter explains the singularities of local connection projects in each country. On the other hand, it focuses on the transnational flows of people, knowledge, and technologies that traversed the isthmus to make local projects possible. Rather than departing from a transnational approach, the discussion of each country individually seeks to demonstrate how transnational flows and exchanges between countries materialized in specific ways at the local level.

As these networking initiatives unfolded, telecommunications markets—owned by state monopolies and controlled by the military in many countries of the region heretofore—were opened for private intervention. In this context, new Internet access providers emerged. This profoundly modified the conditions under which the first networking initiatives operated and the ways in which computer networks evolved in Central America. Chapter 6 discusses how privatizing telecommunications took place in each country of the region as well as privatization's implications for considering development issues.

Finally, Chap. 7 discusses the implications of the evidence presented in the book. It first emphasizes the significance of transnational approaches for examining the history of the Internet. Specifically, it shows how a transnational approach invites a reconsideration of established traditions in the historical analysis of the Internet. Second, it argues that the processes of technological integration examined in the book remain an unfinished project. The book concludes by showing how the history of the Internet in Central America relates to current issues, such as uneven access to computer networks and its implications for understanding the development of the region.

REFERENCES

Bass, J. M., Nicholson, B., & Subrahmanian, E. (2013). A framework using institutional analysis and the capability approach in ICT4D. *Information Technologies & International Development, 9*(1), 19–35.

Bayly, C. A., Beckert, S., Connelly, M., Hofmeyr, I., Kozol, W., & Seed, P. (2006). AHR conversation: On transnational history. *American Historical Review, 111*(5), 1441–1464.

Boltanski, L., & Chiapello, È. (1999). *Le nouvel esprit du capitalisme*. Paris: Gallimard.

Brügger, N., & Milligan, I. (Eds.). (2018). *The SAGE handbook of web history*. London: Sage.

Bull, B. (2005). *Aid, power and privatization: The politics of telecommunication reform in Central America*. Cheltenham, UK: Edward Elgar.

Burt, R. S. (2000). The network entrepreneur. In R. Swedberg (Ed.), *Entrepreneurship: The social science view* (pp. 281–307). Oxford, UK: Oxford University Press.

Castells, M. (1996). *The information age: Economy, society and culture* (Vol. 1: The rise of the network society). Oxford, UK: Blackwell.

CEPAL-BID. (1997). *La integración centroamericana y la institucionalidad regional*. Mexico: CEPAL.

Chan, A. S. (2014). *Networking peripheries: Technological futures and the myth of digital universalism*. Cambridge, MA: MIT Press.

Clavin, P. (2005). Defining transnationalism. *Contemporary European History, 14*(4), 421–439.

Couldry, N. (2012). *Media, society, world: Social theory and digital media practice*. Cambridge, UK: Polity Press.

Davidson, A. (1986). Archaelogy, genealogy, ethics. In D. C. Hoy (Ed.), *Foucault: A critical reader* (pp. 221–233). Oxford, UK: Blackwell.

Gillespie, T. (2007). *Wired shut: Copyright and the shape of digital culture*. Cambridge, MA: MIT Press.

Goggin, G., & McLelland, M. (Eds.). (2017). *The Routledge companion to global internet histories*. London: Routledge.

Hepp, A. (2012). *Cultures of mediatization*. Cambridge, UK: Polity Press.

Hoffmann, B. (2004). *The politics of the internet in third world development*. London: Routledge.

Hughes, T. P. (1983). *Networks of power: Electrification in Western society, 1880–1930*. Baltimore, MD: Johns Hopkins University Press.

Hughes, T. P. (1986). The seamless web: Technology, science, etcetera, etcetera. *Social Studies of Science, 16*(2), 281–292.

Iriye, A. (2002). *Global community: The role of international organizations in the making of the contemporary world*. Berkeley, CA: University of California Press.

Iriye, A., & Saunier, P.-Y. (2009). Introduction: The professor and the madman. In A. Iriye & P.-Y. Saunier (Eds.), *The Palgrave dictionary of transnational history* (pp. xvii–xx). New York: Palgrave Macmillan.

Kaijser, A., van der Vleuten, E., & Högselius, P. (2016). *Europe's infrastructure transition: Economy, war, nature*. New York: Palgrave Macmillan.

Keohane, R., & Nye, J. S. (Eds.). (1972). *Transnational relations and world politics*. Cambridge, MA: Harvard University Press.

Kleine, D. (2013). *Technologies of choice? ICTs, development, and the capabilities approach*. Cambridge, MA: MIT Press.

Kohlrausch, M., & Trischler, H. (2014). *Building Europe on expertise: Innovators, organizers, networkers*. New York: Palgrave Macmillan.

Latour, B. (1991). Technology is society made durable. In J. Law (Ed.), *A sociology of monsters: Essays on power, technology and domination* (pp. 103–131). London: Routledge.

Latour, B. (2005). *Reassembling the social. An introduction to actor-network theory*. Oxford, UK: Oxford University Press.

Law, J. (1987). Technology and heterogeneous engineering: The case of portuguese expansion. In W. E. Bijker, T. P. Hughes, & T. Pinch (Eds.), *The social construction of technological systems: New directions in the sociology and history of technology* (pp. 111–134). Cambridge, MA: MIT Press.

Law, J., & Callon, M. (1988). Engineering and sociology in a military aircraft project: A network analysis of technological change. *Social Problems, 35*(3), 284–297.

Mansell, R. (2001). Digital opportunities and the missing link for developing countries. *Oxford Review of Economic Policy, 17*(2), 282–295.

Mattelart, A. (2000). *Networking the world, 1794–2000*. Minneapolis, MN: University of Minnesota Press.

Medina, E. (2011). *Cybernetic revolutionaries: Technology and politics in Allende's Chile*. Cambridge, MA: MIT Press.

Medina, E., da Costa Marques, I., & Holmes, C. (2014). Introduction: Beyond imported magic. In E. Medina, I. d. C. Marques, & C. Holmes (Eds.), *Beyond imported magic: Essays on science, technology, and society in Latin America* (pp. 1–23). Cambridge, MA: MIT Press.

Morris, A. (2009). *Socio-technical systems in ICT: A comprehensive survey* (Technical report no. DISI-09-054). Trento, Italy: University of Trento.

Penix-Tadsen, P. (2016). *Cultural code: Video games and Latin America*. Cambridge, MA: MIT Press.

Pérez Brignoli, H. (2010). *Breve historia de Centroamérica*. Buenos Aires, Argentina: Alianza.

Russell, A. L. (2017). Hagiography, revisionism & blasphemy in internet histories. *Internet Histories, 1*(1–2), 15–25.

Sánchez Sánchez, R. A. (2009). *The politics of Central American integration*. London: Routledge.

Sánchez-Ancochea, D., & Martí i Puig, S. (2014). *Handbook of Central American governance*. London: Routledge.

Sassen, S. (2017). Digital cultures of use and their infrastructures. In J. Wajcman & N. Dodd (Eds.), *The sociology of speed: Digital, organizational, and social temporalities* (pp. 72–85). Oxford, UK: Oxford University Press.

Schot, J., & Scranton, P. (2014). Making Europe: An introduction to the series. In W. Kaiser & J. Schot (Eds.), *Writing the rules for Europe: Experts, cartels, and international organizations* (pp. ix–xv). New York: Palgrave Macmillan.

Scott, W. R. (2014). *Institutions and organizations* (4th ed.). Thousand Oaks, CA: Sage.

Sen, A. (1999). *Development as freedom*. Oxford, UK: Oxford University Press.

Siles, I. (2008). *Por un sueño en.red.ado. Una historia de Internet en Costa Rica (1990–2005)*. San José, Costa Rica: Editorial de la Universidad de Costa Rica.

Siles, I. (2017a). 25 years of the internet in Central America: An interview with Guy de Téramond. *Internet Histories, 1*(4), 349–358.

Siles, I. (2017b). *Networked selves: Trajectories of blogging in the United States and France*. New York: Peter Lang.

Silva, L., & Figueroa, E. B. (2002). Institutional intervention and the expansion of ICTs in Latin America: The case of Chile. *Information Technology & People, 15*(1), 8–25.

Snyder, S. B. (2011). *Human rights activism and the end of the Cold War: A transnational history of the Helsinki network*. Cambridge, UK: Cambridge University Press.

Takhteyev, Y. (2012). *Coding places: Software practice in a South American city*. Cambridge, MA: MIT Press.

van der Vleuten, E. (2006). Understanding network societies: Two decades of large technical system studies. In E. van der Vleuten & A. Kaijser (Eds.), *Networking Europe: Transnational infrastructures and the shaping of Europe, 1850–2000* (pp. 279–314). Sagamore Beach, MA: Science History Publications/USA.

van der Vleuten, E. (2008). Toward a transnational history of technology: Meanings, promises, pitfalls. *Technology and Culture, 49*(4), 974–994.

van der Vleuten, E., Anastasiadou, I., Lagendijk, V., & Schipper, F. (2007). Europe's system builders: The contested shaping of transnational road, electricity and rail networks. *Contemporary European History, 16*(3), 321–347.

van der Vleuten, E., & Kaijser, A. (Eds.). (2006). *Networking Europe: Transnational infrastructures and the shaping of Europe, 1850–2000*. Sagamore Beach, MA: Science History Publications/USA.

Winner, L. (1980). Do artifacts have politics? *Daedalus, 109*(1), 121–136.

CHAPTER 2

Matters of Central American Integration (1960s–1990s)

Abstract This chapter considers more carefully the context that made it possible for Central American countries to connect to computer networks. It focuses on regional integration initiatives developed in the 1960s (i.e., the creation of the Central American Common Market) and in the 1990s (i.e., the Central American Integration System). It examines the civil war of the 1980s and how peace was reached at the end of the "lost decade." This chapter argues for understanding the mutual configuration between regional integration processes and technological projects. It develops original evidence to show how integration efforts since the 1960s have acquired a technological dimension and how technological projects have also shaped how integration processes were understood.

Keywords Central America • Integration • History • Telecommunications

Connection to computer networks in Central America has to be understood in light of the region's decades of political, economic, social, and technological development. This chapter links development in these areas to the projects for connecting computer networks in the final two decades of the twentieth century, and in that sense, it provides context for the projects analyzed in subsequent chapters. To this end, regional integration initiatives are discussed, specifically those which were conceived in the 60s

© The Author(s) 2020   19
I. Siles, *A Transnational History of the Internet in Central America,*
1985–2000, Palgrave Macmillan Transnational History Series,
https://doi.org/10.1007/978-3-030-48947-2_2

(through the Central American Common Market) and the 90s (by means of the Central American Integration System).

Moreover, a second aim of this chapter is to make evident in practice what previous paragraphs established in theory, namely, how social and technological projects in Central America mutually came into being throughout the latter half of the twentieth century. With this aim, analysis of regional integration (seen by its promotors as a path to development) is articulated with discussion of technological projects. This chapter presents original findings regarding the ways that integrationist efforts starting in the 60s acquired a technological dimension, and how in turn those technological projects shaped understanding of the integration process. In the 60s and 70s, this was carried out by building an analog microwave network that was called the telecommunications "regional artery"; at the end of the 80s and start of the 90s, integration would be sought through "interconnection" initiatives, that is, by implementing computer networks that made it possible for integration to materialize.

The Central American Common Market and the Telecommunications "Regional Artery"

Central American integration has constituted "a constant preoccupation in the region since independence" (Bulmer-Thomas, 1987, p. 177). Torres Rivas (1971) for instance, referred to a "Central American society" united by shared economic and social structure and composition (p. 32) (translated from original source, as are all quotes from Spanish language publications in the present work). Some authors have nuanced the intensity of this supposed tradition, suggesting that the integrationist preoccupation has assumed varying manifestations depending on the historical moment and on each of the region's nations (Granados Chaverri, 1985; Pérez Brignoli, 2017; Sánchez Sánchez, 2009). In other words, not all of the region's countries have sought integration, nor have they wanted the same type of integration, nor have done so in an uninterrupted way. The economic and social asymmetries, as well as the reiterated lack of political will, may also have frustrated these projects' becoming materialized (Rosenberg & Solís, 2007). In practice, beyond the integrationist impetus, if Central America is a society or nation in common, it would be one that has been relatively "divided" (Woodward, 1976). The integrationist project seems perpetually "unfinished" and in an "embryonic state" (Solís, 2000, p. 61). Cerdas (2005) puts it more directly:

> Central America's unionist history has been one of misunderstandings and rivalries, of militarism, war, and violence, of yielding to foreign interests, of criminal tyranny as well as brutal social and racial discrimination, all in clear, vivid contrast with the cherished proclamations of idyllic brotherhood, the endlessly repeated and frustrated dreams of union, and the unionist declarations which are as florid as they are ineffective. (p. 17)

One of the twentieth century's most significant integration efforts dates back to the 60s, specifically to the creation of the Central American Common Market and to the process of industrialization that accompanied it. At the start of the 50s, the Central American States Organization had been created, which was dedicated to promoting integration. The General Treaty of Central American Economic Integration, signed in December 1960 by the region's countries and in 1963 by Costa Rica, established the legal foundations for operating a common market in the region. For a number of authors, the Treaty was the result of integrationist efforts promoted by the Economic Commission for Latin America and the Caribbean (ECLAC) (Bulmer-Thomas, 1987). Others have preferred an interpretation that emphasizes the interference of outside governments (notably that of the United States) which promoted a certain form of integration favorable to their commercial interests (Torres Rivas, 1971). In practice, the Central American Common Market entailed three main dynamics: creating a free trade zone for products originating in the region, conferring fiscal incentives on new industries, and instating various regional organizations (among which stood out a permanent office to execute the Treaty, and a regional development bank [The Central American Bank for Economic Integration]).

The economic and political results of the Central American Common Market have been widely documented. Nevertheless, the decade's integrationist impetus had important implications in terms of infrastructure and telecommunications, which have received relatively less academic attention. Integration was sought in a material sense by means of infrastructure that would allow the region's countries to communicate with one another better. For this reason, the project included a component that was called "physical integration," which included transportation, communications, electricity, and ports (SIECA, 1973).

The greatest advances were seen in the areas of transportation and communications. Work was carried out on development of a "Central American road network," a system of fourteen highways extending more than 5000 kilometers, which was intended to make free trade feasible with the very best transport conditions, both within the region's countries and between them.

In the field of telecommunications, operators experienced important changes during the decade, by means of which they acquired autonomous stature or significant institutional reforms that expanded their range of action. The work to establish a communications system with regional scope can be dated back to the end of the 50s. The operators in each country envisioned this infrastructure as a "necessity" and linked it to two phenomena: in first place, a concept of telecommunications as a direct route to social, economic, political and cultural development, and in second place, the operating requirements of the Central American Common Market (COMTELCA, 1967, p. 1).

The major referent for this network's design was the growing use of telephones in the region, and the need to create conditions that would improve telephone communications between the region's countries. Promoters of the initiative asserted that "expansion [of telephone use] in Central America over the coming years was projected at 14% annually, which implies an increase much higher than that of more developed countries" (COMTELCA, 1967, p. 1). Between 1959 and 1969, the number of telephones in the region's countries doubled, and in the case of Costa Rica, the number even tripled. Besides telephone communications, it was expected that this network would permit the interregional development of other technologies such as "telegraph, telex, radio broadcasting, and television" (COMTELCA, 1967, p. 79).

An initial meeting took place in 1962 to give shape to a concrete proposal for installing what was at first called the "Regional Telecommunications Network of Central America and Panama" and then came to be known as the "regional artery" of communications. Following these conversations, state telecommunications operators submitted a request for funding to the United Nations in order to carry out a study of the status of telecommunications in the region.

The United Nation's Special Fund selected the International Bank of Reconstruction and Economic Development, which contracted a group of French engineers (the group was known as the "French Mission") to do a regional study between 1963 and 1964. The conclusion of this study was to recommend "that a regional telecommunications artery be installed […] in the Isthmus and put into operation" and that a regional enterprise be established to administer and oversee the network (COMTELCA, 1967, p. 2). The details of said installation and operation were worked out in a series of meetings held from 1963 to 1966. Following recommendations derived from the study, the original proposal sought ways to find a regional enterprise that would be in charge of overseeing the operations. However, this plan had to be discarded due to the lack of support on the part of some of the region's countries. The main objection was a sense of loss of local sovereignty over telecommunications on the basis of regional integration.

A step toward greater consensus was taken in early 1966 with the signing of a treaty between El Salvador and Honduras to establish a microwave link between both countries. On the basis of this experience, El Salvador, Guatemala, Honduras and Nicaragua signed a "Telecommunications Treaty" in the same year, with Costa Rica signing it as well in 1967. This treaty was intended to articulate development in telecommunications with the political aspiration of regional integration. Thus, the treaty expressed their "wish to give effective support to the noble ideal of the Central American Union" and to "intensify closer ties and mutual understanding between the signatory nations, these being efforts that must tally with the endeavors of Central American Integration" (COMTELCA, 1967, p. 123). The proposed means of making this political-technological intention materialize in practice was the creation of the Regional Technical Telecommunications Committee (in Spanish, COMTELCA), comprised of the directors of the signatory countries' telecommunications enterprises. The treaty sought to resolve the major objections to the previous proposals, and so it delegated specific responsibilities: while the Regional Technical Telecommunications Committee would be in charge of establishing and operating the regional microwave network between the signatory countries, the local operator in each country would cover the costs of installation and maintenance.

In 1966, the Committee was formed in accordance with those agreements, and preparation was begun that same year to start the bidding process in order to acquire the needed equipment (which consisted of microwave terminal stations to be installed in each nation's capital city, along with the needed repeater or booster stations). Figure 2.1 shows this telecommunications network's more than 1500 kilometers route through a region without internal borders. Inaugurated in 1971, the "regional

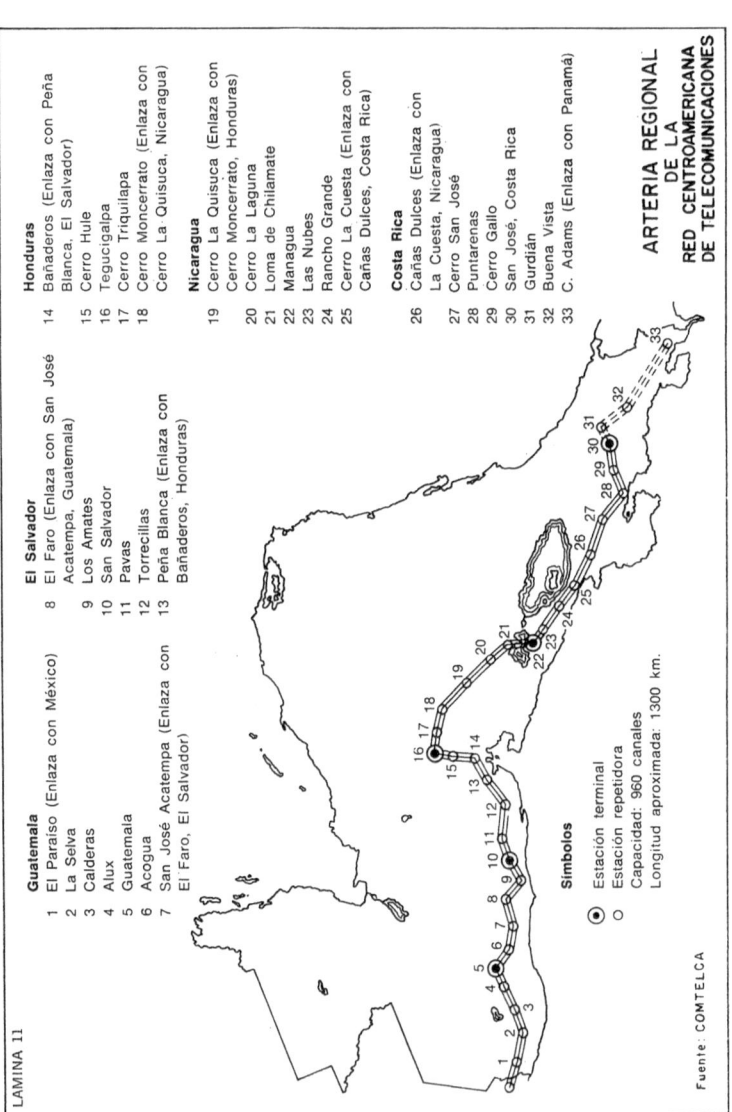

Fig. 2.1 Microwave links between capital cities of Central America (1971). (Source: COMTELCA 1967)
Title: Regional Artery of the Central American Telecommunications Network
Symbols:
- Terminal station
- Repeater or booster station
Capacity: 960 channels
Approximate length: 1300 kms

artery" connected the capital cities of these countries, with access points to the north (Mexico) and south (Panama). As the following decades passed, satellite antennas were installed in each country's territory for further international access to the microwave network. This network molded the integration process in the sense of endowing the very notion of integration itself with a material dimension beyond economic indicators. By means of this network, integration had come into being through antennas and repeater stations whose signals (in the form of telephone calls) crossed the borders of the region's countries.

These "physical integration" projects were among the most important accomplishments of the regional project in the 60s (INTAL, 1983). Advances in integration in the economic and political realms were less clear than those in the telecommunications sector. In 1969, Honduras and El Salvador declared war on one another, which although brief, evidenced profound differences in the region's development.

At the start of the 70s, despite evident modernization in Central America's productive apparatus, the Central American Common Market showed signs of exhaustion. The benefits derived from the market were distributed unevenly among the signatory nations of the treaty. Sánchez Sánchez (2009) points out that, moreover, "the integration process became a casualty of its own decision decision-making system because of governments' competition and the unwillingness to compromise state sovereignty" (pp. 79–80). The discontent arising from economic, political and social deterioration toward the close of the decade would result in a critical situation at the regional level, with the worldwide economic recession as the final trigger (Wiarda, 1984).

CRISIS, PEACE, AND A TANGLED INTEGRATION

The decade of the 80s was marked by "crisis, death, and desolation" (Pérez Brignoli, 2010, p. 211). The integrationist enthusiasm of the 60s, which had peaked with the creation of the Central American Common Market and its telecommunications "regional artery," was destroyed by the reality of a war dedicated to getting rid of the "old order" of the region's dependence and inequality.

The war spread through practically the entire isthmus, except in Costa Rica. Somoza's fall in 1979 ended the dictatorship lasting 46 years which had been backed by the United States in its "backyard." Paraphrasing Pérez Brignoli (2010, p. 186), the Nicaraguan Sandinista revolution pushed the region to the point of conflagration. Throughout the decade, the fight between the Sandinistas and the "contras" grew progressively worse. In El Salvador and Guatemala, these years were marked by military repression against guerrilla forces and social protest movements of both the masses as well as indigenous groups. The toll was thousands of deaths, disappearances, and massive flows of refugees within and outside the region.

This critical situation shrouded the region's countries with a singular geopolitical importance. Following Ronald Reagan's presidential victory (in 1981) the US intervention in Central America intensified in the context of the Cold War and the rise of the region's revolutionary offensive (Smith, 1996). As Rosenberg and Solís (2007) noted, "Washington could not resist the pressure to use Central America as a geopolitical peon in its Cold War chess game" (p. 23). Owing to its geographical location, Honduras turned into a strategic site in the conflict (Rosenberg, Millett, Singer, Weaver, & Shepherd, 1986). A number of US military bases were set up in the country, and its army was re-equipped to fight against the Nicaraguan alliance with the socialist bloc of Cuba and the Soviet Union. In Panama, the decade concluded with a US military operation to arrest General Manuel Antonio Noriega on drug-trafficking charges. Added to this political situation was an economic collapse that proved difficult to reverse and which worsened poverty and inequality. In Costa Rica, a country which had managed to maintain a certain political stability in the midst of the regional maelstrom, this crisis settled in early on and was acutely felt. The confluence of all these factors explains why the 80s are called Central America's "lost decade."

The peace negotiation process (which had materialized in the signing of the Esquipulas II Agreement in 1987) marked the start of a new political context in the region. In the first place, it situated the Central American region as the focus of international cooperation efforts and humanitarian assistance (FLACSO, 2006; Fuentes Knight, 1989; Sánchez Sánchez, 2009). While the isthmus sought an exit from the crisis, it received financial aid from international governments and organizations, at least up to the early 90s. Likewise, the peace negotiation process gave new impetus to regional integration efforts (Sanahuja, 2009). As Sánchez Sánchez (2009) points out, the Esquipulas peace process "meant a reassertion of the importance of the region as a political and economic unit [and] it became the instrument through which the governments would gradually rebuild Central American regionalism" (pp. 122, 129). To be sure, the revitalization of regionalism in the 90s was a wider phenomenon with manifestations in different parts of the world, and characterized by political and economic convergences. Yet, the singularity of this "new regionalism" in Central America "[was] to a large degree, a response to the region's particular conditions, such as its integrationist tradition or the regional dynamic of peace" (Sanahuja, 1998, p. 20). In this context, integration was conceived precisely as a means of pacification after a decade marked by disintegration and fragmentation.

This regionalist project acquired institutional (and legal) status through the Central American Integration System, which reflected a response on the part of Central American governments to the challenges of globalization and the transition to democracy in the region's countries. Approved within what was known as the 1991 Tegucigalpa Protocol and finally implemented on January 1st, 1993, the Central American Integration System proposed a kind of integration that was relatively different from its predecessors of previous decades. For instance, the political plane was extended, as well as the economic one. The Central American Integration System's purpose was to make Central America "a region of peace, freedom, democracy, and development" (ODECA, 1991, p. 195). This aspiration evidenced a clear continuation of the Esquipulas II agreements. At an institutional level, the project included a General Office (based in El Salvador) and a Central American Parliament (set up in Guatemala). As may be intuited from the term "system," the project endeavored to create a joint group of institutions whose vocation was to coordinate regional policies rather than imposing them.

In the framework of this agreement, throughout the decade the signing of a large number of new treaties and conventions to polish the terms of integration was frequent. The Alliance for Sustainable Development, signed in 1994, stipulated a longer-term, more ambitious agenda for the regional integration, and moreover linked it in an intrinsic manner to social and environmental development. The 1995 signing of the Guatemala Protocol reactivated the General Treaty from the 60s. Integration for the economic plane centered on exports as a source of growth more than on industrialization, and on developing complementarity among the region's countries. One of the proposals in that sense was to improve the region's governments power to negotiate with international commercial partners. For that reason, the project has been catalogued as "open regionalism" (CEPAL, 1995b), in contrast with the "introspective scheme" which characterized the Central American Common Market (Bulmer-Thomas, 1998, p. 37).

As an integration project, the Central American Integration System produced mixed results. Though intra-regional commerce grew significantly during the decade, the focus on external competitiveness came at the expense of some internal integration (Sánchez Sánchez, 2009). Solís (2000) points out the lack of political will to endow the Central American Integration System with true capacity to act. Other criticism has centered on the institutional dimension of the integration process (CEPAL-BID, 1997). Thus, Cerdas (2005) condemned gaps between the intentions of organizations designed to favor integration, and the reality that characterized their work. For Cerdas, the integrationist project of the 90s seemed to be marked by "unionist rhetoric" and a "provincial or village mindset" rather than by work conceived to solve the "socio-political and institutional problems of the new integration" (Cerdas, 1998, p. 246). Some measures were agreed to in Panama in 1997 in order to try to remedy the institutional deficiencies of the process. Sanahuja (2009) summarizes the accumulated effect of these combined problems: in practice, the integration process carried out by the Central American Integration System may have been more of an "inter-institutionalized ongoing framework of inter-governmental cooperation" rather than a true integration of a "nature that went beyond national concerns" (p. 46).

In most of the region's countries, the 1989 and 1990 presidential elections were won by conservative political parties. In this way, the neoliberal agenda inspired by the "Washington Consensus" found fertile ground in the region (Williamson, 1990). This normalized implementing legislative measures which favored privatization (such as in education and telecommunications) as well as a significant reduction in the size of the state (Pérez Brignoli, 2010). For some authors, this represented a deeper structural factor than any integrationist project, and it basically brought to a halt any chance of true development and social stability in the region during the 90s (de la Ossa, 1998).

In terms of telecommunications, Central American integration continued on a trajectory similar to the institutional process of regional integration. During the 80s, the Regional Technical Telecommunications Committee sought to maintain the microwave network in order to sustain the flow of telephone communication in the short and long term. Some studies suggest, nonetheless, that at the start of the 90s regional actors in the field of telecommunications viewed the Committee's role as being more reactive than purposeful, given its emphasis on telephone communication (Sáenz & Galeano, 1996). The Central American Integration

System also included a "physical integration" component, although this mainly centered on transportation and energy (CEPAL, 1995a).

In order to foster integration in the area of telecommunications, a technology that had started to peak beginning in the mid-80s was resorted to: computer networks. On the initiative of Roberto Herrera Cáceres, Secretary General of the Central American Integration System, a division (called Information and Network Systems) was created to carry out the mission of exploring the possibility of using computer networks to support and fortify the substantive activities of the very process of integration. The division was headed by Jorge Calvo-Drago, a political scientist interested in matters related to computers.

In 1994, this division implemented the Information Network of the Central American Integration System, which was referred to in Spanish as SICANet. This was an initiative with two major purposes: on one hand, implementing the decisions agreed to within the framework of the Central American Integration System, and on the other hand, facilitating coordination of work carried out among the system's agencies. This process originally involved a "pre-Internet network," as Calvo-Drago describes it (interview with the author, January 22, 2018). Specifically, it used an X.25 network that interconnected the region's countries over the microwave link base ("the regional artery"). By means of e-mail and file-sharing, this network connected different organizations involved in the integration process. (The following chapter describes how the X.25 infrastructure was developed in the region). As the Internet grew to become more firmly established as the standard of computer networks, this division's efforts concentrated on providing access to the network, and on using its possibilities for concrete integration activities: following up on decision-making, implementing regional agreements, communication with the civilian population, and foreign relations. In the words of the initiative's director: "This network provided a centralized teamwork environment that was able to provide connectivity to offices and agencies distributed throughout all of Central America's cities" (Calvo-Drago, 1997, paragraph 19).

The Central American Integration System's projects illustrate a common expectation about the possibilities that computer networks held for augmenting mechanisms of communication and for circulating information. For the Secretary General of the Central American Integration System, integration was pursued as a specific project that could produce certain concrete indicators and results. Beyond the Regional Technical Telecommunications Committee and the Central American Integration

System, integration provided a discourse, logic, and context that favored creating networks—both technological and human—as a path toward development. On one hand, integration permeated connection projects by means of computer networks that flourished with the arrival of the 90s, and on the other hand, the computer networks gave breadth and material expression to the integrationist project.

CONCLUDING REMARKS

Central American history has been marked by recurrent efforts to bring about integration. This chapter examined the major integrationist projects of the twentieth century: the Central American Common Market, and the Central American Integration System. Both projects were manifested in technological development projects and were in turn shaped by them. This chapter sought to make visible the mutual constitution between projects dealing with society (expressed in visions of regional integration) and those of a technological nature (materialized in specific infrastructure). In the 60s, the integrationist project was oriented towards the region's interior and privileged a mainly economic dimension. This resulted in the desire to instate a market with open borders that would facilitate the region's industrialization. A key element of this project was the construction of a microwave network between Central American capital cities which was called the telecommunications "regional artery."

Despite its achievements, the dream of integration ended up buried by the "lost decade" of the 80s. However, a "new regionalism" emerged in the 1990s as part of the pacification process. This project was oriented toward a globalized world and had a more ambitious agenda than the efforts made in the 60s, considering how it tied integration to the region's development. The "new regionalism" materialized in the implementation of "interconnection" initiatives by means of computer networks that made it possible to coordinate the work of institutions that pursued the goal of integration.

Beyond their contributions to integration, networking projects developed in the framework of the Central American Integration System can be interpreted as a metaphor: they expressed a belief which was gradually generalized in that era about the power of computer networks to materialize integration and development in the isthmus. In effect, the project of connecting the region through computer networks transcended the entities in charge of carrying it out in the 90s. The most ambitious proposals for constructing a regional telecommunications infrastructure came from actors outside the institutional integration structure. For these actors, integration provided a compass point of development and computer networks a mechanism to achieve it. For example, state telecommunications operators in each country sought to build a network that functioned over the "regional artery" established at the start of the 70s. Non-governmental organizations also began to work on regional networking projects as a mechanism for social change. Finally, universities sought to come out of academic isolation by means of exchange with colleagues in other parts of the world, made possible by these networks. These are the stories told in the following chapter.

REFERENCES

Bulmer-Thomas, V. (1987). *The political economy of Central America since 1920.* Cambridge, MA: Cambridge University Press.

Bulmer-Thomas, V. (1998). El Mercado Común Centroamericano: Del regionalismo cerrado al regionalismo abierto. In V. Bulmer-Thomas (Ed.), *Centroamérica en reestructuración: Integración regional en Centroamérica* (pp. 21–45). San José, Costa Rica: FLACSO.

Calvo-Drago, J. (1997). *Regional integration of Central American countries and opportunities for internetworking.* Paper presented at the Internet Society (INET) conference, Kuala Lumpur, Malaysia.

CEPAL. (1995a). *Centroamérica: Evolución de la integración económica durante 1994 y avances en los primeros meses de 1995* (LC/MEX/L.283). México: CEPAL.

CEPAL. (1995b). *El regionalismo abierto en América Central: Los desafíos de profundizar y ampliar la integración* (LC/MEX/L.261). México: CEPAL.

CEPAL-BID. (1997). *La integración centroamericana y la institucionalidad regional.* Mexico: CEPAL.

Cerdas, R. (1998). Las instituciones de integración en Centroamérica. In V. Bulmer-Thomas (Ed.), *Centroamérica en reestructuración: Integración regional en Centroamérica* (pp. 245–276). San José, Costa Rica: FLACSO.

Cerdas, R. (2005). *Las instituciones de integración en Centroamérica: De la retórica a la descomposición.* San José, Costa Rica: EUNED.

COMTELCA. (1967). *Proyecto de la Red Regional de Telecomunicaciones Centro América: Estudio de factibilidad. Primera etapa 1970–1980.* San José, Costa Rica: COMTELCA.

de la Ossa, A. (1998). Integración centroamericana y desarrollo social: Los desafíos pendientes. In J. A. Sanahuja & J. Á. Sotillo (Eds.), *Integración y desarrollo en Centroamérica: Más allá del libre comercio* (pp. 95–131). Madrid, Spain: Instituto Universitario de Desarrollo y Cooperación.

FLACSO. (2006). *Los desafíos de Centroamérica desde la perspectiva de las mujeres.* Heredia, Costa Rica: FLACSO.

Fuentes Knight, J. A. (1989). *Desafíos de la integración centroamericana*. San José, Costa Rica: FLACSO.

Granados Chaverri, C. (1985). Hacia una definición de Centroamérica: El peso de los factores geopolíticos. *Anuario de Estudios Centroamericanos, 11*(1), 59–78.

INTAL. (1983). *Evaluación de la integración centroamericana* (No. 163.2). Buenos Aires, Argentina: INTAL.

ODECA. (1991). *Protocolo de Tegucigalpa a la carta de la Organización de Estados Centroamericanos (ODECA)* (pp. 194–205). San Salvador: Organización de Estados Centroamericanos.

Pérez Brignoli, H. (2010). *Breve historia de Centroamérica*. Buenos Aires, Argentina: Alianza.

Pérez Brignoli, H. (2017). *El laberinto centroamericano: Los hilos de la historia*. San José, Costa Rica: CIHAC.

Rosenberg, M. B., Millett, R., Singer, M., Weaver, E., & Shepherd, P. (1986). *Honduras: Pieza clave de la política de Estados Unidos en Centroamérica*. Tegucigalpa, Honduras: Centro de Documentación de Honduras.

Rosenberg, M. B., & Solís, L. G. (2007). *The United States and Central America: Geopolitical realities and regional fragility*. London: Routledge.

Sáenz, M., & Galeano, N. (1996). *Estado de las telecomunicaciones en Centroamérica: Impresiones sobre la situación actual*. San José, Costa Rica: Fundación Acceso.

Sanahuja, J. A. (1998). "Nuevo regionalismo" e integración en Centroamérica, 1990–1997. In J. A. Sanahuja y J. Á. Sotillo (Eds.), *Integración y desarrollo en Centroamérica: Más allá del libre comercio* (pp. 13–58). Madrid: Instituto Universitario de Desarrollo y Cooperación.

Sanahuja, J. A. (2009). Integración regional en América Central, 1990–1997: Los límites del gradualismo. *América Latina Hoy: Revista de Ciencias Sociales, 17*, 43–58.

Sánchez Sánchez, R. A. (2009). *The politics of Central American integration*. London: Routledge.

SIECA. (1973). *El desarrollo integrado de Centroamérica en la presente década* (Integración física) (Vol. 6). Buenos Aires, Argentina: BID/INTAL.

Smith, C. (1996). *Resisting Reagan: The U.S. Central America peace movement*. Chicago: University of Chicago Press.

Solís, L. G. (2000). *Centroamérica 2020: La integración regional y los desafíos de sus relaciones externas* (CA 2020: Working paper no. 3). Hamburg, Germany: Institut für Iberoamerika-Kunde.

Torres Rivas, E. (1971). *Interpretación del desarrollo social centroamericano*. San José, Costa Rica: EDUCA.

Wiarda, H. J. (Ed.). (1984). *Rift and revolution: The central American imbroglio*. Washington, DC: American Enterprise Institute for Social Policy.

Williamson, J. (1990). *Latin American adjustment: How much has happened?* Washington, DC: Institute for International Economics.

Woodward, R. L. (1976). *Central America: A nation divided*. Oxford, UK: Oxford University Press.

CHAPTER 3

The Founding Networks of Central America

Abstract This chapter begins the empirical analysis of the networking experiences and efforts that preceded access to the Internet in the region. The notion of "founding networks" allows for discussion of two important processes. First, the chapter analyzes projects to connect to early computer networks in the region. These projects promoted political visions that materialized in the use of different technologies (i.e., X.25, UUCP, and BITNET). Second, the term "founding" is used to discuss the formation of transnational networks of collaborations among people in several countries of the region. To distinguish them from technological networks, these transnational exchanges were often referred to as "human networks." In Central America, the formation of "human networks" between actors and organizations of the isthmus allowed access to early computer networks.

Keywords BITNET • Computer networks • UUCP • X.25

Over the course of the 70s and 80s, the idea to network computers began to gain traction. This came into being through numerous initiatives with "diverse technical approaches, management philosophies, and purposes" (Abbate, 2000, p. 200). The first computer networking initiatives were carried out in that context. This chapter discusses the establishment of Central America's *founding* networks, in two senses of the expression.

© The Author(s) 2020   33
I. Siles, *A Transnational History of the Internet in Central America,*
1985–2000, Palgrave Macmillan Transnational History Series,
https://doi.org/10.1007/978-3-030-48947-2_3

On the one hand, the region's first computer networking projects are analyzed. These projects promoted political visions that manifested themselves in the use of distinct technologies. For example, the region's state telecommunications operators found an opportunity in X.25 technology to expand the telephone model into a new telecommunications era. During the process of its implementation, they created a regional networking market and infrastructure that would be vital for a future Internet connection. For non-profit organizations (NGOs) and several universities of the isthmus, UUCP technology brought hopes of social change, integration, and development. Finally, in BITNET, academics from the region's universities searched for a solution to the issue of isolated Central American scientists.

On the other hand, the term "founding" is used to discuss the formation of transnational networks between people in various countries who established productive and long-lasting mechanisms of collaboration. In the context of "new regionalism" and integration (examined in the previous chapter), the creation of networking configurations and regional collaboration flourished. One privileged mechanism for implementing integration was the creation of exchange networks with counterparts in various countries of the region. To distinguish them from technological networks, these transnational exchanges were often referred to as "human networks."

This chapter tells the story of how these founding networks interacted in specific projects towards the end of the 80s and the start of the 90s, and how these efforts paved the way for an eventual Internet connection.

RACSAPAC: The Central American X.25 Network

A particular institutional setting in the region's telecommunications field characterized the development of early computer networks. In the 80s, in all of the countries of the Central American isthmus, a state telecommunications company operated under the protection of a monopoly: the National Telecommunications Administration (ANTEL) in El Salvador; the Guatemalan Telecommunications Enterprise (GUATEL); the Honduran Telecommunications Enterprise (HONDUTEL); the Costa Rican Institute of Electricity (ICE) and its subsidiary the Costa Rican Radiographic Institution (RACSA); the National Telecommunications Institute (INTEL) in Panama; and the Nicaraguan Institute of Telecommunications and Postal Services (TELCOR). In the case of

Guatemala and Honduras in particular, their telecommunications companies had close ties to the military and "played key roles in surveillance of the opposition and as income generators for the armed forces" (Bull, 2005, p. 13).

In that institutional framework, one of the first regional computer network interconnection initiatives emerged. The key player in that initiative was the Costa Rican Radiographic Institution, the subsidiary of Costa Rica's state telecommunication operator. Over the course of the 70s, the Costa Rican Institute of Electricity had focused predominantly on offering telephone and electricity utilities. In contrast, its subsidiary—a company whose shares had been acquired in two tracts (in 1964 and 1975) by the institute—had delved into communication technologies like Telex. The existence of two pubic companies in charge of Costa Rican telecommunications is significant, not only because it implied a specialization in services offered, but also because it put the Costa Rican Radiographic Institution, as a subsidiary, in the context of a constant search for business opportunities in order to justify its existence. This permitted a relatively autonomous state agency to focus specifically on the development of Costa Rican telecommunications. At the start of the 80s, the institution was defined as a "telematics" business—a term coined at the end of the 70s by Nora and Minc (1978) to highlight the growing convergence between telecommunications and computer science—and had consolidated its activities in the country's business sector.

Following its first foray into the world of Telex, the Costa Rican Radiographic Institution took a significant turn with the implementation of a new technology in the region: packet-switched public data networks. A first precedent was the acquisition of modems by the subsidiary, which were rented to clients so that they could share resources via terminals located far from a central computer, using the country's telephonic infrastructure. Although this somewhat satisfied local demand, subsidiary corporations from organizations abroad were looking for this kind of service at an international scale. The corporation began to search for a specialized option for those purposes.

To that end, the Costa Rican Radiographic Institution inaugurated two nodes in 1984 that gave access to the international data networks Telenet and Tymnet.[1] These networks were "public" in the sense that, contrary to similar networks installed by private organizations and corporations of the time, they could be accessed by anyone with a computer and a modem who could pay the respective operator's fees. The service was baptized under the name "RACSA-DATOS." An advertisement of the time promised potential clients a "new and advanced system for sending and receiving classified information on a global scale." The company's slogan dictated: "Communicate with us; we'll communicate you with the world." Service subscribers would call a specific number provided by a local operator, which relayed the international link that permitted access to information on other computers. Then, users would pay based on the speed of the connection and the quantity of exchanged data. By the end of 1986, it was estimated that the service had some 300 clients in Costa Rica (Estrada Solano & Rojas Marenco, 1986; Torres, s.f.).

After enabling access nodes to international networks, the second step taken by the Costa Rican Radiographic Institution was the establishment of its own public data network. The project gained a wider scope with the arrival of Marco Antonio Cruz as the company's new chief executive officer (CEO). Cruz came from the planning sector of the Costa Rican Electricity Institute. Within the company, Cruz's arrival (and that of assistant manager Carlos Moreno) was considered a determining factor in the company's expansion into new technological fields during the 80s and 90s. To develop its network, and following the recommendation of the International Telegraph and Telephone Consultative Committee of the International Telecommunication Union, the Costa Rican Radiographic Institution opted for X.25 technology.

The choice of X.25 technology was not merely fortuitous. X.25 had been in development by the International Telecommunication Union since the mid-1970s, and it consisted of a series of communication protocol standards in wide area networks. Because it utilized conventional telephone lines, it was regularly employed for telephone companies' packet-switched networks. Abbate (2000) demonstrates that the X.25's design responded to a political and economic vision of network development, which can be summarized as follows: "The [telecommunication] carriers intended to create a centralized, homogeneous internet system in which network operators controlled network performance" (p. 167). Thus, X.25 protocol made it possible for telecommunications companies to use the existing telephone infrastructure and to maintain access control for each country's computer networks.

The project of creating Central America's first data network implied a significant expense for the Costa Rican Radiographic Institution, to the extent that it required not only the acquisition of equipment to install in Costa Rica (which was purchased from the German company Siemens), but also training for its engineers. Two groups of engineers from the company traveled for several months to Germany in 1986 and 1987, in order to get to know hardware and software aspects related to the network's implementation and operation. After their return, the company prepared itself for launching the new service. In 1987, the company inaugurated its X.25 network, called RACSAPAC. The network consisted of a dozen of nodes distributed in the country's seven provinces. Access to international data networks was extended to include TRT, MCI and ATT, in addition to Tymnet and Telenet. Once the investment had been made, a market had to be created for it. The company immediately launched a search for clients for its new technology. Due to the expense that implementing the network had generated, the institution focused on looking for those clients in corporate and governmental sectors.

In spite of some advances, RACSAPAC's market in Costa Rica was limited. In that context, the company's management considered expanding its market to the rest of Central America. Thus, just a few months after the X.25 network's release in Costa Rica, the Costa Rican Radiographic Institution initiated a series of conversations with regional counterparts to create access nodes to its X.25 network in each country of the region. The efforts were fruitful and, from 1988 to 1990, telecommunications operators from other countries in the region enabled access nodes to RACSAPAC or their own data networks: ANTELPAC in El Salvador, TELEDATOS in Honduras (where two nodes were installed), MAYAPAC in Guatemala y NICAPAC in Nicaragua. The institution's engineers traveled to each of these countries to work on installing nodes and training regional counterparts. To facilitate access to RACSAPAC from these countries, the "regional telecommunications artery" established in the 70s was employed (see previous chapter). In this way, a Central American X.25 network was formed that operated through a microwave link administrated by the Regional Technical Telecommunications Commission.

The selection of this technology allowed for a type of specific interconnection that Abbate (2000) describes with precision: "They [postal, telegraphic, and telephone operators (PTTs)] envisioned that each country would have a single public data network, and that the various public networks would interconnect at national borders" (p. 165). In other words, these efforts were about a centralized integration: each country maintained its autonomy to control public networks from the inside; the network only opened borders to data flow. From that point of view, Central

America was predominantly seen as an economic market for developing an innovative service.

Around the time of this initiative, an emerging computer culture was galvanized in several of the region's countries. One of the main applications available through the network were public databases. The Costa Rican Radiographic Institution (RACSA) had spent its time intensely promoting the use of and access to these databases. These efforts were the result of what the company perceived as an "insufficient boom in service supply" through its public data network; the company intended to maximize its use and, in this way, "stimulate the development of a computer culture necessary to access" the network (RACSA, 1992, p. 2). In Costa Rica, for example, the number of databases rose significantly in the second half of the decade. From 1988 to 1991, that number increased from 48 to 291 (Cerdas, 1992). Some examples of these were the Public Registry databases (1985), those of *La Nación* newspaper (to cover the 1988 presidential elections, the Presidential Summit held in the country in 1989, or the 1991 solar eclipse), those of the Costa Rican Tourism Institute (1989) or of the company itself (RACSA), which was utilized to provide information about its services to clients (1989).

In addition to databases, public data networks provided the context for creating some of the first bulletin board systems (BBS) in the region's countries, which could be accessed remotely by way of a modem and in this way exchange emails, access computer-stored information, and share online files with other people (Pérez, 1996; Siles, 2008).[2] According to engineer Mónico Oyuela, who administrated a BBS hosted by the research and development branch of Honduras's state telecommunications operator, the system had a practical function: to reveal the possibilities for communication of the country's newly born network (interview with the author, November 21, 2017).

Perhaps the region's most prominent BBS was the Midnight Express (*El Expreso de Medianoche*). Created in Costa Rica in 1987 by Carlos Revilla, a young computer enthusiast, the Express owed its name to the time when it could be used. Given that access to the Express required a telephone line, Revilla was authorized at work to utilize one line after work hours and before the start of the following work day. That way, the Midnight Express operated from 5 p.m. to 8 a.m. In addition to email, the BBS gave access to computer exchange forums as well as to data on varied topics.

The Express experienced an unprecedented rise in popularity within the emerging Central American computer culture. Against its success, the Costa Rican Radiographic Institution investigated the possibility of establishing a similar service that could compete with the Express (Serrano, 1994; Siles, 2008). In 1990, Revilla signed a contract with this company to offer the BBS together. By then, the Express offered an impressive amount of services and applications: free software, email capable of transferring files, forums on a range of diverse topics, a shopping section, online games, and search for the country's databases. Over the next several years, the BBS would also offer services such as faxing and the possibility to connect to the Internet. The profits obtained by per-minute use of the BBS were divided: 66% for the company and 33% for Revilla.

Finally, X.25 public data networks allowed for the development of other types of projects and initiatives. One notable example was the ambitious plan developed by the Omar Dengo Foundation in 1990 to provide computer network access as educational tools in Costa Rica's public schools (Fonseca, 1991). At the start of the new decade, the Costa Rican Radiographic Institution inaugurated a satellite access service that concluded a decade of technological expansion and ended up with a solid infrastructure that would be fundamental for the regional interconnection projects of the coming years.

UUCP: A TECHNOLOGY FOR PEACE AND DEVELOPMENT

Inasmuch as local telecommunications operators installed an infrastructure that permitted regional interconnection through public data networks (i.e., the X.25 network), several initiatives based out of universities and NGOs would use this infrastructure for various purposes. These efforts came to be in early connection initiatives by way of two specific technologies: UUCP and BITNET.

The set of protocols and programs called Unix-to-Unix-Copy (UUCP) was instrumental in the initial interconnection of Central America through computational networks. Developed in the mid-70s at AT&T Bell Laboratories, UCCP gave access to an email network and computer file exchange by way of the operating system Unix (Hauben & Hauben, 1997). UUCP operated under the technique of store and forward: information was sent to the nearest node, where it was stored and then sent to the next node, and so on until it reached its final destination. In addition to a relatively low cost (given that it did not require any additional infrastructure) and the possibility to operate on any machine with Unix, UUCP had acquired a political identity that Paloque-Berges (2017) captures accurately: "the UUCP protocol was encompassed with a hacker aura and do-it-yourself technical practice [...] Unix networks were based on an open, distributed, and collaborative model, keen on including new users from many professional backgrounds" (pp. 156–157). These factors, combined with the creation of networking configurations as a mechanism of regional integration, would give UUCP a fertile trajectory in the Central American isthmus. This trajectory was fed by two principal sources: on the one hand, a group of NGOs, and, on the other hand, universities linked for the most part to a Central American integrationist network.

Nicarao and Progressive Networks for Social Change

In the second half of the 80s, an interest surged in several Central American NGOs to connect to both email networks and social activist information-exchange networks. A few early examples at a global scale included GreenNet, developed from 1985 in the U.K. and dedicated to environmental discussions; PeaceNet, a network of peace activists established in the same year and located in the U.S.; and EcoNet, an ecological network based in the U.S., created in 1986. In 1987, PeaceNet merged with EcoNet to form the Institute for Global Communications (IGC), with its headquarters in San Francisco, California. This institute served as a network of UUCP nodes dedicated to discussing human rights and the environment. By 1990, it joined six more networks to form the Association for Progressive Communications (APC), a worldwide network described by one of its founders as "an electronic tool [to] create a world of cooperation [...] [and construct] a future for the planet that works a lot better than the present [...] a telecommunications service closely linked to citizen action" (Mitra & Miller, 1988, p. 2). Behind these projects lied a fundamental premise: the networking of progressive organizations had the "power" to make a difference in the world (Mitra & Miller, 1988; Murphy, 2000, p. 29). In this way, because of its technological characteristics, UUCP was seen as a radically different political project compared to other alternatives in computer networking history (Murphy, 2004).

In Nicaragua, a group of researchers associated with an NGO called Regional Coordinator of Economic and Social Investigation (CRIES) had utilized Costa Rica's public data network to exchange information with PeaceNet since 1987 (Boschmann & Richards, 1990). In the following months, a more formal and direct connection was sought. The Nicaraguan NGO signed an agreement with the Institute for Global Communications and, in June 1989, installed an access node to email through UUCP. The node, called Nicarao, was connected in this way to PeaceNet. This made Nicarao the first APC node in Latin America.[3] The name of the node referred to the pre-Columbian group who settled in what is known today as Nicaragua and to the mythical chief who may have governed the southeast region of the country at the start of the sixteenth century (Fowler, 1985).

Nicarao's start date was not fortuitous. The coming presidential elections in Nicaragua, to be carried out in February 1990, gave the project a certain sense of urgency. Brian Coan, an IGC collaborator in San Francisco who traveled to Managua in May 1989 with a computer and modems for node installation, and who stayed for eight months to participate in its development, recalls:

> There was a lot of United States propaganda to discredit or sabotage the election in some way or another. Information was needed from people who were in Nicaragua. The story that we wanted to get out was what these people were doing and what the situation on the ground was like. The elections were the driving force and what made things happen so quickly. (Interview with the author, December 21, 2017)

In that sense, the link with APC was important not only for the incoming information but also because it signified a means for communicating news about the region's countries during the political and economic reconstruction following the peace negotiation. Nicaragua's node gave the Institute for Global Communications information "from the ground" in an area of great geopolitical value. Likewise, the connection with the Institute provided regional NGOs a series of technological tools for communicating with other parts of the world.

The Nicaraguan node connected twice a day through a long-distance call to San Francisco, to exchange emails (a technology that had been reserved up until that moment for academic entities and private organizations). Another significant application were "conferences," described by a leader of the Association for Progressive Communications as "online discussion groups based on the Usenet model [...] [that] allow discussion moderators to control who has access to reading, writing, and accessing the online space" (Hackenthal, 2000, p. 8). The technical team at the

Institute for Global Communications worked to develop applications and technologies that would make it possible to utilize the existing infrastructure and to reduce the high cost of network data transmission.

As an NGO dedicated to the investigation of social studies, the Regional Coordinator of Economic and Social Investigation was linked to various organizations in the rest of the Central American isthmus. The exchanges between Nicarao's administrators and their regional counterparts facilitated the use of the node in the other Central American countries. In other words, the network of contacts that this NGO held in the region helped to spread its use. To access Nicarao, users in other countries employed the regional X.25 network. In that way, Nicarao became one of the first providers of technologies such as email in the region's countries. In 1990, according to Boschmann and Richards (1990), the node was used by one hundred people.

Interest in networking issues in Nicaragua gained a fundamental ally with the arrival of German engineer Cornelio Hopmann. Hopmann had worked at a state-run research center for computer science and applied mathematics in Germany since the beginning of the 70s. In that sense, he had already connected to "what would soon be known as Internet" since 1983, not without certain reservations on the part of the military mindset that had permeated the network's development (interview with the author, July 22, 2017). Motivated by the possibility to help establish a field in computer science and software engineering, Hopmann traveled to Nicaragua at the end of the year and settled permanently in the country in 1985. Hopmann explains:

> After doing reconnaissance [in 1983], Nicaragua's connectivity at that time was an item on my agenda [...] The launch of the PeaceNet Nicarao node was [an advancement] [but] from the beginning the objective for the next step was to again have what I had had in 1983: [Internet] connection with UUNET. (Interview with the author, July 22, 2017)

Established as a nonprofit organization in 1987, UUNET became the first commercial Internet service provider in the United States. Its roots trace back precisely to the use of UUCP and the efforts to connect it to the Internet (Malik, 2003; Swisher, 1996).[4] In Central America, UUNET was the privileged intermediary between the first computer network nodes (notably UUCP) and the Internet. In March 1989 (several months before Nicarao's installation), Hopmann established an access node to UUNET from the National University of Engineering, where he had directed the creation of the Computational Engineering department since his arrival to Nicaragua. The node at the university (UNINIC) was connected to the APC node in Nicaragua, but then sought a direct link to UUNET (Arce & Hopmann, 2002). In doing so, UNINIC became the UUNET representative in Nicaragua.

A Node Called Huracán

Another project with a similar objective to Nicarao's was established in 1990 by Theodore ("Ted") Hope, an American computer engineer and graduate of the Georgia Institute of Technology, who resided in the region from an early age. Hope had worked partially for the Regional Coordinator of Economic and Social Investigation in Nicaragua, where he had seen Nicarao's operation and possibilities. In 1990, Hope moved from Nicaragua to Costa Rica. Thanks to the intermediation of Vincezo ("Enzo") Puliatti, a regional official of the United Nations Development Programme (UNDP) (whose role will be discussed in the following section), Hope got in contact with the Superior Central American University Council (CSUCA), a confederation of Central American universities based out of San José, Costa Rica. With the support of the Council, Hope launched a computer networking project called Huracán. The project was originally financed by UNDP, the Canadian International Development Agency, and the University of Ottawa. To baptize the project, Hope chose a name that made reference to Mesoamerican culture, stemming from a Nahuatl word referring to a Mayan deity. Since 1988, previous to Huracán's development, the Superior Central American University Council had gained access to PeaceNet and attempted to activate email access through this network. However, the project failed due to the cost of the link and several access issues (Richards, 1992). Huracán offered a more permanent and less costly solution to these problems.

Huracán fit perfectly into the political vision of the Council to encourage the integration of Central American universities through technology. Over the course of the 80s, the Council initiated a process of institutional transformation oriented more toward interuniversity collaboration rather than research itself per se. Against that backdrop, one of its main working premises was that "new information and communications technology [constituted] a revitalizing element for regional exchange and integration" (Boschmann & Richards, 1990, p. 151). To this end, a series of projects with support from international organizations had been developed. Among these projects were access to a database (that contained information regarding Council universities and research projects within them), a software program (Microisis), and a teleconference system that connected the seven universities that formed the Council and the University of Ottawa, which had developed the technology. In 1992, Edgardo Richards, Chief Information Officer, described this teleconference infrastructure in the following way:

> [A system of] interactive audiographic communication [that] utilize[d] an electronic board, a monitor, a sound and microphone system, and a voice and data [modem] in each center. A small central switch was responsible for routing the signals, making the [simultaneous] interaction [between universities] possible. (Richards, 1992, p. 382)

In order to implement this combination of projects, a regional user network was necessary. Since the mid-80s, and as part of a growing tendency in the region to establish work networks, CSUCA had created the Central American University Network for Scientific Information (REDCSUCA). Richards expanded on this: "Networks were formed by the people. We were able to engage and unite people who had a genuine interest and a genuine vocation for [Central American] development and integration" (interview with the author, November 16, 2017). On the basis of those efforts, Richards undertook establishing contacts in Central American universities who could act as users of the teleconference system, which favored the subsequent adoption of Huracán.

Technologically-speaking, Huracán utilized UUCP and the X.25 network which connected the region's countries to "promote the use of non-commercial CSUCA-connected computer networks in Latin America and the Caribbean at a relatively low cost for users" (Siles, 2008, pp. 71–72). This vision was crucial for Hope and the Superior Central American University Council, who preferred a low-cost development of the network that would facilitate "reaching a critical mass [of users] sooner rather than later" (Hope, interview with the author, March 17, 2006). Specifically, Huracán gave access to email, Usenet newsgroups, and several databases. Each local call cost approximately US$0.03 to US$0.05 per minute, depending on the country.

To exchange emails with Internet users, Huracán connected to the Internet three times a day by way of UUNET. Hope had discovered

UUNET during his master's in computer science, which he finalized in Georgia in 1988. In turn, thanks to this relationship with UUNET, Hope obtained the top-level domain administration (TLD) for Costa Rica (.cr) in 1990. Hope explains:

> UUNET was fundamental for the pre-Internet development of the Internet. They told [us]: "We'll register the .cr domain name, create the DNS record so that whichever email in the entire world sent to Huracan.cr arrives to us at UUNET, waits in line and you pay for the calls. [...] You pay for the link and we'll worry about the rest on our side, you don't have to pay for a router on this side of Virginia." They never charged us a cent. (Interview with the author, August 01, 2017)

To reduce costs, they resorted to a common practice of the decade (but one which was opposed by the state telecommunications company): the call-back. The trick operated in this way: from Costa Rica, people called a U.S. operator. After a few rings, communication was cut off, which made charging for the call impossible. Immediately, the U.S. operator returned the call to a Costa Rican number and transferred it to another telephone number, whereby it was connected to the network. Because the call was initiated from the United States (and not from Costa Rica), the cost of the international call was reduced considerably.[5]

Toward a Central American UUCP "Backbone"

From 1990 to 1995, Huracán was a major force in the development of UUCP and the transnational flow of technology in the region. In Guatemala, the project found an important promoter in Luis Furlán, a physicist and electrical engineer who directed a center for computer science studies at Guatemala's Universidad del Valle. Furlán was looking for a way to maintain contact with his peers once he returned to Guatemala, after studying in the United States. A coincidence connected Hope and Furlán. Hope had arrived to Guatemala when he was two years old and completed part of his secondary studies at a high school which shared its campus with the Guatemalan university. Both met when Hope was studying computer science courses at the university's campus. After the creation of Huracán, Hope thought of Furlán as "the ideal counterpart" for the project (Hope, interview with the author, August 01, 2017).

In this regard, Furlán recalls:

[Hope] told me: "Why don't give our UUCP protocol technology a try?" And he told me about the Huracán project. So we said, "Well, let's give it a try." [...] We got a modem and started to try it out and, indeed, we achieved the connection between Guatemala and Costa Rica [in 1992], under the Huracán project. And that was our first major step toward the Internet. (Interview with the author, August 10, 2017)

Motivated by Hope, Furlán asked John Postel, coordinator of the Internet Assigned Numbers Authority (IANA), for a .edu domain to manage activities associated with the UUCP node in Guatemala. According to Furlán, Postel wrote to inform him that, as he was the first in the country to make the request, he could also receive the administration of the top-level domain. In that way, the TLD (.gt) administration was turned over to Universidad del Valle in 1992. Just like Nicarao, Huracán's regional launch served to communicate information that had not found outlets by other means. Pasch and Valdés (1997) recall, for example: "During the 1993 coup d'etat [in Guatemala], when the media was forcefully silenced, we shared news reports received via email and through huracan [sic] newsgroups with faculty and students at our university [Francisco Marroquín]" (p. 11).

The UUCP expanded throughout the isthmus thanks to the network of contacts established by the Superior Central American University Council to bring about technologically-led integration efforts in its partner universities. In other words, this computer network inserted itself into the network of collaborations and exchanges that preceded it. Richards describes the required negotiation process for installing new nodes: "I would go to [every university's] Vice-president of Research to raise the issue of Huracán, and I would say to them, 'Who am I speaking with?' That was how I set up the network in every country" (interview with the author, November 16, 2017). Subsequently, Hope traveled to each country to establish the link to the Costa Rican node. In that way, Huracán found partners in Panama, Honduras and El Salvador (as will be discussed further on). Víctor Barragán, staff worker of the Graduate School at the University of Panama, to whom access to Huracán was delegated, aptly summarizes the significance of this system for the country:

During the years when the project developed, in addition to the United States army personnel stationed in the formal Panama Canal Zone and several banks and companies, Huracán was the only computer networking connection option in Panama [...] [It was] a mode of communication that permitted almost 100 of its national users and foreign residents in the country [Panama] to send and receive messages [...] quickly, safely, and cheaply worldwide. (Interview with the author, March 24, 2006)

In Honduras, after conversations between Richards and the National Autonomous University of Honduras, access to Huracán was assigned to the library system. In this way, engineer Eduardo Pleitez became Huracán's Honduran counterpart. The project also captured the attention of Gustavo Pérez, professor at the Honduran university's Physics Department. On a visit to Honduras, Richards explained to Pérez the procedures for connecting to Huracán, which was used during the following years. Thanks to this experience, Pérez would later become an important driving force in the country's Internet connection (see Chap. 5).

Despite international advances, Huracán faced some internal financial challenges. By the end of 1991, the Superior Central American University Council experienced an institutional crisis and, in January 1992, Hope transferred the project to the Nahual Foundation, an NGO in San José, Costa Rica. During 1993, subsidies ran out and Hope started to charge the system's users a certain amount in order to finance the project. By 1995 private Internet service providers began to emerge in various countries of the region. This changed the meaning that projects like Huracán and Nicarao have inasmuch as they expanded ways to access computer networks at increasingly lower costs. In that context, Hope announced Huracán's closing in 1995 by way of an email to users. According to Hope, by that date Huracán had almost 2500 users in the region (interview with the author, March 17, 2006). Huracán was shut down permanently on May 31, 1995.

By the start of the 90s, the nodes of Nicaragua (Nicarao) and Costa Rica (Huracán) were points of regional interconnection and means of information flow about Central America in the years that marked the end of the war. Given that the use of UUCP had been consolidated in every country of the region, some users discussed the possibility of establishing a "Central American UUCP backbone" that would give it more viability and better connection conditions (Ortega, 1993). However, according to Pasch and Valdés (1997), the rise of other computer networks diluted the interest in this possibility, notably the connection to BITNET in Costa Rica.

BITNET: OVERCOMING ACADEMIC ISOLATION

In addition to UUCP, the other network that played a significant role in the Central American region's early interconnection was BITNET. This network was created in 1981 and, over the first year of its life, connected 25 computation centers in the United States, including those of Princeton, Columbia, and North Carolina (Grier & Campbell, 2000). Similar to UUCP, this network was created with computational purposes, but it was rapidly adapted by its users to facilitate communication processes (Abbate, 2000, p. 202). BITNET also employed the store and forward technique to circulate information. Unlike UUCP, it operated solely between IBM machines.

BITNET grew significantly from 1987 to 1990. Its popularity in American academic spaces grew as its access costs lowered. According to Grier and Campbell (2000), the boom of BITNET during the 80s can also be explained by the success experienced by its electronic mailing list system, called Listserv. BITNET reached its peak in 1990, when it came to connect 3500 nodes distributed in 46 countries. Half of those nodes were in the United States, while the others were distributed among users of the European Academic Research Network, NetNorth in Canada, GulfNet in the Middle East, as well as in Japan (AsiaNet), Latin America, and Africa. The example of countries closest to Central America, notably Mexico, consolidated interest in connecting to this network from Costa Rica (Gutiérrez, 2017; Siles, 2008).

The First Central American Node: UCRVM2

The first connection to this network in the Central American isthmus was established in 1990 in the University of Costa Rica with the installation of the UCRVM2 node. It was connected through a satellite link between San José and Florida, U.S., where the closest node to the network was located (Siles, 2012). BITNET's history in Central America is transnational in its own right. In other words, the Central American network's beginning can only be explained by the flow of people and technologies inside the region and with other parts of the world.

The connection was led by a Costa Rican-French physicist named Guy de Téramond, who had used networks like BITNET and the Internet thanks to academic experiences in the United States and France. De Téramond returned to Costa Rica at the end of the 80s with the desire to have a local access point and to forge better communications relations with colleagues abroad (Siles, 2017). His main motivation was to overcome Central American isolation by way of links to counterparts around the world and access to relevant information. De Téramond wrote at the end of the 80s in the journal *Physics Today:* "[the] link [to these networks] can solve one of the worst scientific problems in third-world countries: isolation" (1990, p. 95). Access to BITNET was seen as a solution to this challenge in academia.

The decision to connect to BITNET was considered practically "obvious," given the availability of two IBM machines at UCR (de Téramond, 1994, p. 67). De Téramond expands on this:

> The Internet was not so known by physicists at that time, afterward everything changed very quickly. I also discovered that in the [UCR] Computer Center no one knew about Unix, so we had to start with the equipment we already had and what people knew. There were already various terminals on campus connected to the [IBM] 4381. [...] Additionally, the required [BITNET] link had lower capacity than what was needed for the Internet [...] It was suitable for the support that we had at that time, and it was the same that was being done in Latin America, both in Chile as well as in Mexico. (Interview with the author, November 02, 2005)

De Téramond's efforts were not the first in the country. Dr. Claudio Gutiérrez, who had served as president of the University of Costa Rica from 1974 to 1981 and who had worked in the early 80s as professor at the University of Delaware, led various unsuccessful initiatives to achieve connection. The same had occurred with the attempt to connect to VNET, IBM's corporate network, led by computer scientist Francisco Mata, who had completed his master's studies at Case Western Reserve University. These failed experiences transformed Gutiérrez into a self-designed "frustrated predecessor" (interview with the author, November 29, 2005). At the same time, they had an instructive effect: they permitted interested actors, particularly de Téramond, to devise a proposal that could address the difficulties faced by previous plans (notably mechanisms to get sufficient funds and to scale up the project in the short-term).

In January 1990, during a sabbatical leave from the University of Delaware that he spent in Costa Rica, Gutiérrez contacted de Téramond to express his worries about the challenges of connecting the country to computer networks (de Téramond, 1994). With Gutiérrez's support, de

Téramond intensified his efforts to achieve connection. Using Law's (1987) concept, the work method that de Téramond adopted could be labeled "heterogeneous engineering": it involved the establishment of a network of allies between a wide variety of actors and organizations that could give the project financial viability and provide it with legitimate institutional and political legitimacy in the country (Siles, 2012). Within UCR, de Téramond obtained the support of its respective authorities and became the spokesperson for actors interested in achieving connection, mainly academics who had used these networks while they studied abroad. In addition, de Téramond established a small work team composed of four computer science professionals affiliated with the UCR Computer Center.

Outside of the university, de Téramond consolidated an alliance with interested actors in other academic institutions of the country, including the National Council for Scientific and Technological Investigation, and established conversations with the necessary organizations to establish the BITNET link, notably the state telecommunications company and IBM's local representatives. The project benefitted from what Téramond called "a series of circumstances that worked in our favor" (interview with the author, June 23, 2017). Specifically, de Téramond referred to the launch of the world's first digital communications satellite by the Pan American Satellite corporation (PanAmSat).

The Pan American Satellite corporation was created in 1984 by Rene Anselmo, co-founder of the television chain known today as Univision, to provide satellite communication services between Latin America and the United States (Rubin, Neuman, & Phillips, 1994). Together with other companies, the corporation requested approval in 1984 from the Federal Communications Commission (FCC) to operate a private satellite communication system and thereby break up the International Telecommunications Satellite Organization (Intelsat) monopoly (LaCroix, 1993). Intelsat was a "a non-profit international organization run on commercial principles" that, by the end of the 80s, operated in tandem with telephone and telecommunications providers in more than 115 countries in the world (Gershon, 1990, p. 249). In 1990, it was estimated that Intelsat handled "70% of the world's telephone calls and virtually all international television transmission" (Gershon, 1990, p. 249). FCC approved the PanAmSat operation in 1988, the same year in which it launched its first satellite: PAS-1. The company rapidly expanded its scope of action to grow into a global alternative to Intelsat.

The link to the Pan American Satellite corporation in Costa Rica came about through an initiative by the Costa Rican Radiographic Institution, the subsidiary company to the state telecommunications operator, which explored options to install a satellite antenna. The company had experienced some pressure from transnational companies to facilitate satellite communications in the country's duty-free zones. In that sense, the Pan American Satellite corporation provided a better option for data transmission than Intelsat, given the emphasis of this organization on analog data transmission and its connection to telephone service providers. Alberto Bermúdez, an electrical engineer hired by the Costa Rican Radiographic Institution to participate in the satellite project (because of his experience in developing satellite stations at the Costa Rican Electricity Institute) and who years later would become the company's CEO, recalls:

> The first thing this gentleman, Marco [Cruz, then CEO] told me [in 1986] was: "I need you to go to Comsat [Communications Satellite Corporation], because [the Costa Rican Electricity Institute station in] Tarbaca can only transmit analog telephone service, but we need to transmit over digital channels." The star product [would be] a 64-Kbps channel to rent to international corporations such as banana corporations, shipping enterprises, and duty-free companies. (Interview with the author, March 15, 2018)

The first Costa Rican Radiographic Institution satellite station was established in this way through the Pan American Satellite corporation. When de Téramond and his task force approached the Costa Rican Radiographic Institution to discuss conditions and possibilities to connect to BITNET (and eventually the Internet), the company put them in contact with the Pan American corporation. The association with this corporation proved key for Costa Rican network connection plans. Specifically, it provided a less costly alternative to establish the satellite link than Intelsat. The Pan American Satellite corporation allowed BITNET promoters to "have at their disposal double the transmission flow for half the cost of the line" offered by Intelsat (Cerdas, de Téramond, & Gutiérrez, 1990, p. 684). On a structural level, this permitted de Téramond to establish a crucial alliance with Brien Morgan, PanAmSat engineer, who would be decisive factor in the region's Internet connection process (see Chap. 5).

In March 1990, Max Cerdas (Computer Science Commission member of the National Council for Scientific and Technological Investigation), de Téramond, and Gutiérrez presented a plan to connect to BITNET at the

Space Conference of the Americas, in San José, Costa Rica, that system-
atized the efforts and conversations carried out up to that moment. The
plan's title is telling in numerous ways: "International electronic connection for Central American scientists: A BITNET node installation project
in Costa Rica" (Cerdas et al., 1990). The document began with a description of BITNET, explained the possibilities offered by the network for
academic work, specified the procedures and requirements necessary for
achieving the connection from the University of Costa Rica, enumerated
implementation costs, and concluded with the intention to achieve
"regional extension of the project" (p. 685). As concrete evidence of the
intentions, it cited a dozen of Central American universities interested in
the proposal. To conclude, the authors stated:

> The countries of Central America, Costa Rica in particular, have a unique
> opportunity to equip their scientists with some of the better development
> conditions that their counterparts in more developed countries have access
> to [...] The interactive use of the network [...] helps to a large extent to
> eliminate the isolation of scientists from small institutions by giving them
> instruments that only larger institutions can maintain. (p. 687)

Among the presentation's audience members was Orlando Morales,
Costa Rican Minister of Science and Technology. Morales showed interest
in the project and, after a meeting with de Téramond, became a governmental ally. The funds to establish the connection arrived thanks to the
network of allies established by de Téramond's work team. In 1989, the
Inter-American Development Bank had approved a loan to the National
Council for Scientific and Technological Investigation (CONICIT) to
finance the National Science and Technology Project. With this loan, the
council had proposed the development of a National System of Scientific
and Technological Information that took into account the creation of a
network of seven Specialized Information Centers. The connection to
BITNET turned out to be strategic for the Council in the sense that it
made it possible to connect to the seven chosen information centers and
thereby equip them with the network's available resources. With that aim
in mind, and thanks to the support of political and academic actors, the
Council approved the use of $45,000 for the BITNET connection project
(Cerdas & Porras, 2017; Siles, 2008).

Furthermore, with a donation from IBM negotiated by de Téramond,
an IBM machine was acquired, in which protocol converters were installed

which would make access to BITNET possible from any computer in the country or Central American region (using, for example, the X.25 network). Against that backdrop, connection to BITNET was finally achieved in November 1990. The magazine *Prociencia*, an official publication of the National Council for Scientific and Technological Investigation, announced the news in its November 1990 edition:

> Costa Rican scientists will have, as of this year, the possibility to exchange experiences with groups of investigators from the best American, Japanese, Australian, and European universities thanks to the country's link to the BITNET Network [...] Costa Rica will be the first Central American country to have this information service [...] It is anticipated that the second stage of Costa Rican connection will extend throughout the entire isthmus. (1990, p. 5)

The launch of the UCRVM2 node permitted the arrival of a very attractive application for the pioneer user community: email. The task force provided users with an introductory guide to the network, and gave conferences and training courses about its operation in the university community.

Meanwhile, de Téramond investigated the requirements for establishing Internet connection. One of the first steps he took was to request from the Internet Assigned Numbers Authority the top-level domain administration for Costa Rica (.cr), which had been conceded to the Superior Central American University Council thanks to Hope's efforts. The domain was assigned to de Téramond in September 1990 (two months before connecting to the Internet). Nevertheless, de Téramond decided to have the National Academy of Sciences take charge of handling the TLD, rather than the University of Costa Rica or the state telecommunications institution. The decision, which caused controversy within UCR, was defended by Téramond as a mechanism to avoid a sites management monopoly.

From Costa Rica to Panama: UTPVM1

Connection to BITNET quickly acquired a regional dimension through the flow of people between Costa Rica and Panama, as well as the role of IBM as international actor with a presence in the region. Victor López, Vicedean of Research at the Technological University of Panama at that time, and one of the participants in the project in that country, remembers:

The connection to BITNET emerged because of the interest of one of IBM's staff members, José González. He was a student at Santa María La Antigua University. And on a trip that IBM took to Costa Rica, he visited the University of Costa Rica and saw the whole BITNET operation. He returned believing that the same could be done and that this was going to be his thesis project. (Interview with the author, July 07, 2017)

To establish this connection, it was decided that an IBM 4361 computer would be used from the School of Engineering and Computation Systems at the Technological University of Panama. González and López traveled once again to the University of Costa Rica to seek advice about the connection's technical and administrative aspects. After a series of conversations with representatives from other universities, IBM, and Panama's National Institute of Telecommunications, the UTPVM1 node connected to the Costa Rican BITNET node in 1992. By April 1993, a study estimated that the node in Costa Rica was utilized by more than 1700 regional users (including El Salvador, Guatemala, Nicaragua, and Panama, in addition to Costa Rica).

Concluding Remarks

Early computer networking captured a variety of hopes for a Central America that was emerging little by little from crisis. Whether it was in academia, corporations, or nongovernmental organizations, access to these networks was celebrated as an opportunity for development and "a gateway to progress" (Hopmann, 1998, par. 2). Telecommunication operators saw in the X.25 network a way to expand their operation model—inspired by the telephone network—into the world of data. To find clients for their public networks, they supported creating projects such as public database access. Early computation aficionados gave life and purpose to these networks through systems like BBSs. These systems galvanized the emerging Central American computer culture. The X.25 network propelled by telecommunications operators was crucial for the regional development of other computer networks established toward the end of the 80s and at the start of the new decade.

The UUCP had an important trajectory in nongovernmental organizations and some universities of the isthmus. With a specific political identity, it was seen as an ideal mechanism for promoting social change and facilitating regional development. Finally, academics who had used computer networks while studying or carrying out research abroad, returned to initiate local crusades in order to connect to these networks from Central America. They saw in technologies such as BITNET an opportunity to overcome academic isolation and maintain active relationships with peers in other parts of the world; in short, to advance the field of science.

The importance of these networking projects transcends its strictly technological dimension. For some time, early networks provided an international outlet of information about the isthmus in the years following peace negotiation processes. Moreover, these initiatives installed networks of transnational exchanges that would last more than the use of several technologies. The collaborative networks forged by means of these projects did not only outlive UUCP and BITNET themselves, but also, Internet connection would be achieved based on these collaborative mechanisms and flows of knowledge and technologies that constituted these early exchange networks (often called "human networks" at that time, to distinguish them from computer networks [Labelle, 1995; Silvio, 1992]).

These initiatives also had symbolic importance, in the sense that they demonstrated the connection's feasibility and normalized the use of computer networks and applications like email in Central America. In the words of Yadira Mena, collaborator to the project at the National University of Engineering in Nicaragua, these initiatives removed the "esoteric" aspect from the networks (interview with the author, July 14, 2017). In Honduras, as Gustavo Pérez states, "Huracán made us wonder. Those of us who had experienced email, we had this idea that a better means of communication could exist, and at a much cheaper price, and we would be willing to fight for its installation" (interview with the author, December 04, 2017). Julio Lezcano, who participated in the link to BITNET in Panama, states similarly: "BITNET was the trigger, the gunpowder" (interview with the author, July 18, 2017). The struggle and the boom—often together—would arrive just months later with Internet connection.

NOTES

1. According to Campbell-Kelly and Aspray (1996), the networks Telenet (1975) and Tymnet (1979) were created by vendors of time-sharing computers, in order to reposition themselves in light of the email boom and the early consolidation of computer networks.

2. Based on information collected from various dissertations on the topic, Siles (2008) provided a list of at least 17 BBSs created in Costa Rica by the late 80s and early 90s: Pro Medical Computer Science Association (utilized by medical science professionals from Hospital México), Asopride (Association of Private Developmental Organizations), Byteline, Caribbean, CR Online, Crxzit, Edenia BBS, Midnight Express, Geg Chu, Habitat!, Jac, Lobo, Megaline, PC-Online, Tritón and UNEPNET (Acosta et al., 1992; Sanabria, 1996; Serrano, 1994). Most experienced fleeting success.

3. APC nodes in Latin America included Alternex (Brazil), BolNet (Bolivia), Chasque (Uruguay), Equinex (Ecuador) and Huracán (Costa Rica).

4. In 2000, Tim O'Reilly noted with respect to UUNET: "At the time [1987], it could have been called a UUSP, for "UUCP Service Provider"; TCP/IP didn't become the biggest part of the business till years later. UUCP lives on as the UU in UUnet's name" (2000, par. 9). UUNET distributed O'Reilly's 1986 book about UUCP and Usenet among its clients at no cost.

5. Based on media reports, Hoffmann (2004) states that, by 1999, this practice represented a total of 22% of long-distance calls to the United States from Costa Rica.

REFERENCES

Abbate, J. (2000). *Inventing the internet.* Cambridge, MA: MIT Press.

Acosta, R., Arrieta, J. G., Berrocal, G., Corrales, E., Del Valle, R., & Lorenzo, J. C. (1992). *Bases de datos públicas: Un modelo para Costa Rica (Tomo I).* Memoria del seminario de graduación para optar por el grado de Licenciatura en Ciencias de la Computación e Informática, Universidad de Costa Rica, San José, Costa Rica.

Arce, M. E., & Hopmann, C. (2002). *eReadiness and requirements fundamentals and objectives for the CDG of Nicaragua.* Unpublished work.

Boschmann, I., & Richards, E. (1990). Servicios y proyectos telemáticos en América Central. *Estudios Sociales Centroamericanos, 53,* 151–162.

Bull, B. (2005). *Aid, power and privatization: The politics of telecommunication reform in Central America.* Cheltenham, UK: Edward Elgar.

Campbell-Kelly, M., & Aspray, W. (1996). *Computer: A history of the information machine.* New York: Basic Books, Harper Collins.

Cerdas, M. (1992). *La oferta latinoamericana de bases de datos.* Paper presented at the "Desarrollo de las bases de datos a través de las redes públicas" conference, San José, Costa Rica.

Cerdas, M., de Téramond, G., & Gutiérrez, C. (1990, March 12–16). *Conexión electrónica internacional para los científicos centroamericanos: Proyecto de instalación de un nodo BITNET en Costa Rica.* Paper presented at the Conferencia Espacial de las Américas, San José, Costa Rica.

Cerdas, M., & Porras, V. (2017). Aportes de CONICIT a la conectividad de internet. *Boletín de Ciencia y Tecnología del CONICIT,* no. 171. Retrieved from http://www.conicit.go.cr/prensa/boletincyt/boletines_cyt/bol_171/Aportes_CONICIT_Internet.aspx

de Téramond, G. (1990). Seconding solutions for third world science. *Physics Today, 43*(8), 95.

de Téramond, G. (1994). Interconexión de Costa Rica a las grandes redes de investigación Bitnet e Internet. In *Idearo de la ciencia y la tecnología: hacia el nuevo milenio* (pp. 61–86). San José, Costa Rica: Ministerio de Ciencia y Tecnología.

Estrada Solano, M., & Rojas Marenco, M. A. (1986). *Red pública de datos por conmutación de paquetes.* Trabajo final de graduación presentado como requisito para optar al grado de Licenciatura en Ingeniería Eléctrica, Universidad de Costa Rica.

Fonseca, C. (1991). *Computadoras en la escuela pública costarricense: La puesta en marcha de una decisión.* San José, Costa Rica: Fundación Omar Dengo.

Fowler, W. R. (1985). Ethnohistoric sources on the Pipil-Nicarao of Central America: A critical analysis. *Ethnohistory, 32*(1), 37–62.

Gershon, R. A. (1990). Global cooperation in an era of deregulation. *Telecommunications Policy, 14*(3), 249–259.

Grier, D. A., & Campbell, M. (2000). A social history of Bitnet and Listserv, 1985–1991. *IEEE Annals of the History of Computing, 22*(2), 32–41.

Gutiérrez, F. (2017). The evolution of the internet in México (1986–2016). In G. Goggin & M. McLelland (Eds.), *The Routledge companion to global internet histories* (pp. 105–121). London: Routledge.

Hackenthal, S. (2000). Message from the APC chair: Ten years after its founding, APC continues to promote the internet as a tool for social change. In APC (Ed.), *APC annual report 2000* (pp. 8–9). San Francisco: APC.

Hauben, M., & Hauben, R. (1997). *Netizens: On the history and impact of usenet and the internet.* Hoboken, NJ: Wiley-IEEE Computer Society Press.

Hoffmann, B. (2004). *The politics of the internet in third world development.* London: Routledge.

Hopmann, C. (1998, March 23). *Internet en Nicaragua: La historia de las oportunidades desaprovechadas.* Retrieved from https://interred.wordpress.com/1988/03/23/internet-en-nicaragua-la-historia-de-las-oportunidades-desaprovechadas/

Labelle, R. (1995). Knowledge and information resources for local and traditional natural resource users. *Yale Bulletin Series, 98,* 176–194.

LaCroix, R. A. (1993). Developments in international satellite communications in the international space year. *CommLaw Conspectus, 1*(1), 99–108.

Law, J. (1987). Technology and heterogeneous engineering: The case of portuguese expansion. In W. E. Bijker, T. P. Hughes, & T. Pinch (Eds.), *The social construction of technological systems: New directions in the sociology and history of technology* (pp. 111–134). Cambridge, MA: MIT Press.

Malik, O. (2003). *Broadbandits: Inside the $750 billion telecom heist.* Hoboken, NJ: Wiley.

Mitra, A., & Miller, D. L. (1988). *Why the association for progressive communications is different.* Paper presented at the 38th annual conference of the International Communication Association, New Orleans, Louisiana.

Murphy, B. M. (2000). The founding of APC: Coincidences and logical steps in global civil society networking. In APC (Ed.), *APC annual report 2000* (pp. 28–30). San Francisco: APC.

Murphy, B. M. (2004). Propagating alternative journalism through social movement cyberspace: The appropriation of computer networks for alternative media development. In R. Eglash, J. L. Croissant, G. Di Chiro, & R. Fouché (Eds.), *Appropriating technology: Vernacular science and social power* (pp. 163–180). Minneapolis, MN: University of Minnesota Press.

Nora, S., & Minc, A. (1978). *L'informatisation de la société.* Paris: La Documentation Française.

O'Reilly, T. (2000, May 9). Lessons from the layoffs at Linuxcare. *O'Reilly Linux Dev Center,* from http://www.linuxdevcenter.com/pub/a/linux/2000/05/09/lessons.html

Ortega, T. (1993). *Propuesta backbone UUCP Centroamericano.* Paper presented at the Seminario Centroamericano para Especialistas en Información, Guatemala, Central America.

Paloque-Berges, C. (2017). Mapping a French internet experience: A decade of Unix networks cooperation (1983–1993). In G. Goggin & M. McLelland (Eds.), *The Routledge companion to global internet histories* (pp. 153–170). London: Routledge.

Pasch, G., & Valdés, C. (1997, February 3–5). *The dawn of the internet era in Guatemala.* Paper presented at the International Federation of Information Processing, Florianopolis, Brazil.

Pérez, M. G. (1996, April 12). Internet en Honduras. *Asuntos Electrónicos,* from http://www.angelfire.com/ca5/mas/electronica/elect016.html

Prociencia. (1990). Costa Rica ingresa a la red BITNET. *Prociencia, XIV*(81), 5.

RACSA. (1992). *Experiencias en el diseño y puesta en marcha de bases de datos públicas.* Paper presented at the "Desarrollo de las bases de datos a través de las redes públicas" conference, San José, Costa Rica.

Richards, E. (1992). Nuevas tecnologías e integración académica en América Central: Experiencia de la Red Universitaria Centroamericana de Información Científica (REDCSUCA). In J. Silvio (Ed.), *Calidad, tecnología y globalización en la educación superior latinoamericana* (pp. 379–388). Caracas, Venezuela: CRESALC-UNESCO.

Rubin, P. A., Neuman, M. E., & Phillips, W. F. (1994). *The PanAmSat satellite system*. Paper presented at the AIAA 15th international communications satellite systems conference, San Diego, CA.

Sanabria, G. A. (1996). *MIVAH BBS, sistema de información en línea sobre vivienda y asentamientos humanos*. Proyecto de graduación para optar al grado de Licenciatura en Computación e informática, Universidad de Costa Rica.

Serrano, M. (1994). *Operación de una base pública de mensajes y archivos basada en protocolo X.25*. Tesis sometida como requisito parcial para optar por el grado de bachiller en Ingeniería Eléctrica, Universidad de Costa Rica.

Siles, I. (2008). *Por un sueño en.red.ado. Una historia de Internet en Costa Rica (1990–2005)*. San José, Costa Rica: Editorial de la Universidad de Costa Rica.

Siles, I. (2012). Establishing the internet in Costa Rica: Co-optation and the closure of technological controversies. *The Information Society, 28*(1), 13–23.

Siles, I. (2017). 25 years of the internet in Central America: An interview with Guy de Téramond. *Internet Histories, 1*(4), 349–358.

Silvio, J. (Ed.). (1992). *Calidad, tecnología y globalización en la educación superior latinoamericana*. Caracas, Venezuela: CRESALC-UNESCO.

Swisher, K. (1996, May 6). Anticipating the internet. *The Washington Post*, from https://www.washingtonpost.com/archive/business/1996/05/06/anticipating-the-internet/ebef84f0-e7d9-4340-96ef-b57ae949da40/?utm_term=.38ca69bea261

Torres, M. (s.f.). *El desarrollo de los servicios radiográficos en Costa Rica*. San José, Costa Rica: Museo Histórico y Tecnológico, Instituto Costarricense de Electricidad.

CHAPTER 4

An Internet for the Global South

Abstract This chapter examines the "regimes of alliances" between a variety of organizations that were formed to enable technological access points to the Internet in Central American. The chapter discusses how organizations such as National Science Foundation, the Organization of American States, and the United Nations Development Program mobilized their political leverage to promote the connection to the Internet in the region. The chapter thus highlights the centrality of international organizations in how Central America connected to the network. The role of these organizations is summarized in three major ways: (a) articulating networks of collaborations between actors of different nature and origin; (b) obtaining and distributing funds to establish Internet access points; and (c) negotiating with governments and state telecom operators to endow networking initiatives with political traction.

Keywords Development • Latin America • Internet • National Science Foundation • Organization of America States • United Nations

The efforts to connect to early computer networks, described in the previous chapter, acquired greater importance and broader scope through a series of transnational initiatives developed by international organizations. Insomuch as the Internet gained stability and spread throughout academia in the United States, these initiatives were oriented toward facilitating its

I. Siles, *A Transnational History of the Internet in Central America, 1985–2000*, Palgrave Macmillan Transnational History Series, https://doi.org/10.1007/978-3-030-48947-2_4

access and development in regions such as Latin America. The purpose of these projects was to turn the Internet into a truly international network and thereby establish it as the computer networking standard.

This chapter examines the body of networks established among various international entities that made this Internet internationalization process possible. By the way in which these networks were created, they can be defined as "regime[s] of alignment" or alliances, that is to say, heterogeneous networks of technologies, institutions and discursive justifications (in this case, pertaining to visions for development), from which fund distribution systems were created (Gillespie, 2007). I analyze institutional projects that gave a broader scope to efforts that had delivered the X.25, UUCP and BITNET connections. In that sense, creating these networks was a driving force in connecting the Central American isthmus to the Internet.

Hereafter, three concrete initiatives are discussed. First, a National Science Foundation (NSF) project to facilitate network access outside of the United States, which more than 12 Latin American countries would benefit from. Second, an initiative by the Organization of American States (OAS) to facilitate Internet connection to its member countries. NSF and OAS became allies to that end. And, finally, a collection of efforts generated by the United Nations Development Programme (UNDP) emphasized the value of computational networks as a vehicle for social development. These projects would help assemble regional actors interested in the Internet and partially finance the links from Central America.

In discussing these initiatives, this chapter makes clear the importance of international organizations in establishing transnational networks of people, knowledge, and technologies that were decisive in the Central American region's Internet connection. The chapter analyzes the transnational action capacity of these organizations, which can be summarized in three ways: (a) the creation of networks among a wide range of actors; (b) fundraising and fund distribution to achieve Internet connection; and (c) negotiations at the governmental and institutional level to give political traction to networking initiatives. In analyzing the role of these organizations, this chapter also takes into consideration the varied political conceptions associated with the Internet and the controversies derived from them.

NSFNET OUTSIDE OF THE UNITED STATES

During the 1980s, the Internet left the military domains in which it had originally been conceived and was gradually adopted in academic environments in the United States. The role of the National Science Foundation in this process has been amply documented (Abbate, 2000; Hafner & Lyon, 1998; Press, 1996). The creation of its NSFNET network was a decisive factor in this sense. NSF's intention was to construct an internet (or a network composed of networks) more than a network in and of itself (Abbate, 2000, p. 192). This strategy had two components: connecting to existing United States supercomputer centers and developing a backbone that would make help connect regional networks and supercomputer centers. According to Abbate, "in the fall of 1985 about 2000 computers had access to the Internet; by the end of 1987 there were almost 30,000, and by October of 1989 the number had grown to 159,000" (2000, p. 186). Nevertheless, less well-known than the gradual development of NSFNET in the United States are the efforts made in that era to facilitate its access abroad, most notably in Latin America.

The role of Steve Goldstein and his direct superior at NSF, Stephen Wolff, was key in that sense. During the 1990s, as program director at the National Science Foundation, Goldstein conducted a series of initiatives with the aim, in his own words, "to spread Internet around the world" (interview with the author, November 30, 2017). Goldstein arrived to the computer science division of the Directorate for Computer and Information Science at NSF in 1989, after decades of work at the MITRE Corporation, during which he collaborated with NASA on the establishment of a computer network connected with Internet protocols.

The projects to facilitate access to NSFNET from outside the United States emerged as a response to what NSF perceived as "a growing international necessity" to connect to the network (Goldstein, interview with the author, November 30, 2017). In the case of Latin America, the Organization of American States and NSF convened an initial Latin American networking workshop in July 1989, in San José, Costa Rica, where the following people participated: Steve Goldstein and NSF's Harold Stolberg; the founder of CSNET, Lawrence ("Larry") Landweber (University of Wisconsin-Madison); Glenn Ricart (University of Maryland); Eric Marler (IBM); Enzo Puliatti (UN Development Programme); and representatives from the region's countries interested in computer networking, among others. In assessing the activity, Goldstein

(1991, p. 32) noted that the cost of domestic and international connections was seen as an obstacle.

In January 1990, according to Abbate (2000), "there were 250 non-US networks attached to the NSFNET, more than 20 percent of the total number of networks" (p. 201). In that context, in 1990 NSF issued a solicitation (NSF 09-69, "International Connections to NSFNET") to "consolidate the management and engineering of connections between the U.S. research and education communities and similar communities abroad" (Goldstein, 1995, p. 1). The project had originally been designed to strengthen the connection of two specific networks—considered strategic by the foundation—to NSFNET: INRIA (France) and NORDUnet (Nordic countries) (Lehtisalo, 2005). Following some considerations, connection to other countries through the initiative was approved (Goldstein, 1991). The cooperative agreement was granted to the company Sprint under the name "International Connections Management" (ICM), valid from 1991 to 1996. In this way, Sprint became a key ally to the NSF project. In becoming involved, the company could "gain experience with the Internet," adds Goldstein (interview with the author, November 30, 2017). Due to regulation, NSF could not use the project's available money to make payments in other countries. Instead, the project constituted alliances with actors whose participation would facilitate international connection. In that sense, an agreement was reached with the Pan American Satellite corporation that permitted offering a preferential tariff to any country that used the company's 64 kbps-satellite link to connect to NSFNET.

The mission "to spread Internet around the world" acquired several forms, including support of grassroots initiatives. By the end of the 80s, networking engineer Randy Bush had worked on offering practically-focused technical training in different parts of the world, notably southern Africa. Goldstein found common ground with Bush's initiative and, in late 1991, helped finance Bush's trip to Peru to configure a UUCP connection. With NSF funding, Bush formalized his project in 1992 through the creation of the Network Startup Research Center (NSRC), a nonprofit organization based out of the University of Oregon dedicated to promoting computer networking initiatives in developing countries.

Under the "regime of alliances" (Gillespie, 2007) established by way of the International Connections Management agreement—which linked the National Science Foundation, PanAmSat, Sprint, and the Organization of American States (whose participation will be explained in the following section)—a plan was designed that would have major implications for Latin American connection. NSF proposed facilitating and partially subsidizing an Internet access point in Homestead, Florida, a strategic place for Latin American connection. Using a fiber optic link provided by Sprint, this node in Homestead would connect to the NSFNET gateway in Washington. This offered a new connection possibility to the expanding Internet backbone. Moreover, it would facilitate interconnection between Latin American countries and the PanAmSat satellite, which could connect without having to pass through United States territory. This point of presence in Homestead was implemented in late 1992.

Within the framework of the International Connections Management project, more than 25 countries connected to the Internet, including a dozen in Latin America (Press, 1996). This made Goldstein into "*our* [Latin American] contact in the United States," as called by José Soriano, Peruvian Internet promoter, on a website dedicated to documenting the region's Internet histories (Soriano, 2007). Parallel to efforts to make the network available outside of the United States, NSF worked to facilitate the network's privatization at the start of the next decade (Abbate, 2010). Insomuch as conditions to privatize the network were created, the userbase continued to grow, and developers of other computer networks adopted TCP/IP protocols, the Internet "absorbed" the body of existing networks (including UUCP and BITNET) and, at the same time, the content of these networks began migrating to the Internet (Abbate, 2000, p. 205).[1] It was in that context that institutional efforts emerged to facilitate Internet connection to Latin American countries. One of those projects—and a key actor of the regime of alliances that constituted the International Connections Management project—was an OAS initiative called the Hemisphere-Wide Inter-University Scientific and Technological Information Network (RedHUCyT).

RedHUCyT: Internet and Academic Development in Latin America

The project of the Hemisphere-Wide Inter-University Scientific and Technological Information Network (RedHUCyT) was created by the Organization of American States (OAS) in 1991. The project's director was a Mexican doctor of mathematics named Saúl Hahn, who, before joining the OAS in 1987 as Principal Specialist, had worked as professor of mathematics at the Center for Research and Advanced Studies at the National Polytechnic Institute, and as consultant for the IBM Scientific Center, both in Mexico. Hahn considered the project's creation a natural extension of the OAS's role as international body in the region's scientific progress:

> As a scientist, one wishes to communicate with colleagues from other areas. Within the OAS, we always had projects related to information exchange: cassettes, magnetic disks, and databases. Back then the idea of establishing these types of [computer] networks in Latin America was very appealing. (Interview with the author, July 7, 2017)

The project was a "network" in the sense that it sought to unite various actors and organizations as well as academic institutions, governmental entities, telecommunications companies and private corporations, to promote Internet connection in Latin America (Hahn, 1996, 1999, 2000). Hahn explains the creation of the project thusly: "Given the advances that were happening in the United States and in some of the region's countries, we were very interested in creating a network or in supporting various efforts [that already existed in that sense]. In this way, we presented several projects and were able to obtain some external funding" (interview with the author, July 7, 2017).

The project envisioned the Internet as a means to strengthen academic exchange and thus advance science in the Latin American region. The principal objective was to establish "a scientific and technological network for exchanges between professors, researchers and different universities of the OAS member states" (Hahn, 1995, p. 2). Hahn emphasized the possibility of sending and receiving information, and the capacity to access "interactive services" (like email, long-distance computer databases, supercomputer resources, experiments and simulations), as features of the networks that would make forming "cooperative participation" possible in the fields of science and technology (Hahn, 1996, p. 53).

From its beginnings, the project became the Latin American counterpart of the National Science Foundation's International Connections Management project (described in the previous section). NSF's Steve Goldstein explains the practical dynamics of the relationship between both parties: "[Hahn persuaded] national authorities to permit the connections and paid for the necessary equipment for the ground station with RedHUCyT funds. I paid the circuit link in the U.S. [which was] the most

I could do" (interview with the author, November 30, 2017). This alliance with OAS permitted the National Science Foundation to reach its objective of extending the Internet's reach beyond the United States, specifically in Latin America. The foundation, in turn, through initiatives like International Connections Management, offered the Hemisphere-Wide Inter-University Scientific and Technological Information Network project the opportunity to cheapen connection costs, which represented a significant problem for the region's countries.

OAS's transnational action capacity was reflected in many ways. One of the first and most significant was the call for actors from the organization's member countries to activities where networks of exchange and collaboration could be established. These activities had a "networking effect": they made visible the efforts, individuals, and teams that pursued common interests and helped join efforts among them. In June 1991, the International Networking Conference (INET) was held in Copenhagen, where representatives from 11 Latin American countries (including Costa Rica and Nicaragua) attended. Months later, in October 1991, OAS and the Brazilian Council for Scientific and Technological Development organized the "First Interamerican Networking Workshop" in Rio de Janeiro, an event with significant repercussions for the Central American region. (Representatives from Costa Rica and Nicaragua once again attended this workshop). In the worlds of Hahn:

> At this meeting we had around 130 participants. NSF financed the participation of researchers, professors, technicians from the United States. We [the OAS and the Brazilian government] [financed] a large part of the [participants from] Latin America. There were difficult discussions, but it was very useful because we learned what the region's situation was like. (Interview with the author, July 7, 2017)

As part of that regional connection analysis, the principal challenges that needed to be faced to connect to the Internet became very clear: the high costs of a satellite link to the United States and the lack of personnel trained in Internet protocols. An advanced student of industrial engineering, Teresa Ortega, who attended the workshop and would later lead the Nicaraguan connection project, recalls: "We were a bunch of Latin Americans trying to organize ourselves: What could be done? Who would be the first to get connected? With what funding? [...] There we started to have more Latin American contact. For me it was a discovery" (interview with the author, July 30, 2017). The meeting concluded by participants signing a collaboration agreement.

As can be inferred from Hahn's words, obtaining collaboration agreements was not easy. Goldstein highlights the role of Tadao Takahashi, the host of the Rio workshop, in helping transform common interests into specific agreements. In order to give continuity to the work done at the Brazil reunion and to strengthen the formation of collaboration networks, the Hemisphere-Wide Inter-University Scientific and Technological Information Network co-financed annual conferences in countries such as Mexico, Argentina, Venezuela, Peru, and Chile, and helped several collaborators of local projects in Latin America to participate in the various INET conferences organized by Internet Society in the following years (notably in Kobe, San Francisco, Hawaii, and Geneva).

Several plans discussed during the Rio conference for dealing with these challenges marked a major turning point for the Central American connection projects. It was at this meeting that de Téramond discovered the National Science Foundation project to facilitate a fiber optic link between Homestead and the NSFNET gateway in Washington. This possibility provided a new direction for the work of the Costa Rican representatives present at the conference. As a result, the original plan to connect the country to the Internet through SURAnet was abandoned—the Internet network in the southeastern United States (de Téramond, 1994; Siles, 2008). De Téramond recalls another solution, discussed at this meeting, but never implemented: renting a satellite transponder to cover the Latin American region which would allow for a relatively cheaper price for the region's countries. The failed plan shows that the early solutions of Internet access were essentially characterized by regional thinking.

One of the main results of the efforts at these conferences was the formation of alliances which had a multiplier effect in the region. From then on, de Téramond and Hahn built a work relationship oriented toward enabling Central American network access. The details of a concrete connection plan for various Central American countries were thoroughly discussed at a meeting between de Téramond and Hahn held in Washington, D.C. in July 1992, a few months following the Rio de Janeiro conference. In the activity's report, de Téramond indicated:

> Architectural aspects of systems, equipment, as well as political and organizational matters were discussed in detail. It is agreed that terrestrial microwaves from the Regional Technical Telecommunications Commission will

be used to implement the first phase of an ambitious communications project in Central America. (de Téramond & Brenes, 1994, p. 7)

Once again it becomes clear that a regional perspective was present even before the first dedicated Internet connection in Central America was established. De Téramond explains how the project's ideal allies were chosen during that meeting:

> We arrived at the conclusion that the region's universities had to be directly supported, since there were other options, including the NGOs [...] What we perceived was that those options were not going to have the necessary conditions to [scale up] these projects. In contrast, the universities had their own computer centers, engineers, technicians, and students: it was an investment in something that was much more stable that had much higher growth potential. (Interview with the author, June 23, 2017)

This decision expresses, once again, the link established by de Téramond and Hahn between computer networks and the region's scientific progress.

Another way in which the Organization of American States channeled its transnational work capacity was by distributing funds according to the specific needs of different areas of the region: Central America, the Caribbean, and South America. The Hemisphere-Wide Inter-University Scientific and Technological Information Network project (RedHUCyT) worked with actors from each part of Latin America to establish Internet connection based on each region's needs and possibilities. This aid was primarily supplied through economic resources to finance a satellite link (though, in most cases, local countries should have financed permanent access once connection was established), support of existing infrastructure growth (including donations of satellite antennas to improve connection), and training seminars for local project participants which allowed for "improvements in abilities, shared knowledge, and network management training" (Hahn, 2000, par. 3). Given its vision of the Internet as a way to promote substantial scientific work in the region, RedHUCyT was oriented primarily toward universities and academic institutions, which were seen as ideal spaces to scale up the initiatives in the long term (Siles, 2017).

Finally, in addition to building networks and distributing funds, the OAS expressed its transnational action capacity through negotiations with governments and local telecommunications operators to obtain permits and facilitate connection. In the action plan from the first Summit of the

Americas (held in Miami in 1994), the OAS had incorporated a suggestion directed to the region's government to facilitate Internet access of "major universities, libraries, hospitals and government agencies [...] building on the work of the OAS Hemisphere-Wide Inter-University Scientific and Technological Information Network" (OAS, 1994, p. 10) As will be discussed in the next chapter, OAS aid would be crucial for carrying out this suggestion in countries of the isthmus where Internet access was viewed with caution by local telecommunications companies.

UNDP: The Internet as the Path Toward Development

Parallel to the OAS's work, the United Nations Development Programme (UNDP) also made efforts to strengthen regional connection. The programme's projects were motivated by a general understanding of communications technology, and the Internet in particular, as a path to development. This understanding was based on two aspects of computer networks: on one hand, the possibility to obtain and publish relevant information for each country and, on the other, the capacity to establish long-distance collaboration with other individuals. This dual portrayal was extensively documented in writings produced by the United Nations Development Programme in the early and mid-1990s. One of these documents stated:

> The Internet [...] has already promoted a process of "information democratization" by allowing governments, citizens, civil society groups, non-governmental organizations (NGOs) and institutions to capture, publish and distribute information and knowledge resources—including indigenous knowledge and local-language content ... This, in turn, has made feasible the potential for increased cooperation and collaboration between countries to address regional and global development issues. (SNDP, n.d., p. 1)

From this perspective, the ultimate goal was not access *to* technology but the development that could be developed *by* this access. In short, the Internet was an enabling tool for development. Inspired by this vision, various United Nations Development Programme efforts were implemented at the start of the decade.

A Latin American Telecommunications Network

One initiative was coordinated by Enzo Puliatti, a political scientist who worked for the UN Development Programme since 1986. Puliatti became familiar with the topic of computer networks in an orientation offered by the UN for its new employees: "[As] part of the course, they showed us remote database access, email and things like that, and the potential that this could have if it were brought to developing countries impressed me" (interview with the author, November 02, 2017). Since then, he utilized the United Nations Development Programme's transnational action capacity to support local connection initiatives in Latin America. The project's premise was an understanding of the Internet as a tool for social change. As an early description of the project put it, promoters thought "the use of emerging inexpensive technologies that can be easily assimilated by developing countries could provide new models of development" (Yeung & Puliatti, 1992, p. 22). Puliatti explains: "[We viewed] the Internet's development as a revolutionary process, not so much in the political sense, [but] in the sense that it could have an impact on the countries' development and that in that way it had to be promoted in any way possible" (interview with the author, December 02, 2017).

Based on alliances with actors working on expanding network access outside the U.S. (including UUNET and Larry Landweber's work team at the University of Wisconsin), Puliatti's project brought about economic resources so that several people could participate in conferences and training courses as well as the creation of work alliances among interested groups (Yeung & Puliatti, 1992). For example, Puliatti helped finance around 15 trips for Latin Americans to the 1992 International Networking Conference in Kobe, Japan. For Puliatti, the Internet's expansion required finding a network of promoters in the region and equipping them with the necessary conditions to establish a link. "The idea," states Puliatti, "was not only to form networks from a technological perspective; it was to motivate and form groups to create a critical mass in each country and thus connect countries in any way possible as fast as possible" (interview with the author, November 02, 2017). In several countries some equipment (hardware and software) was acquired to achieve connection to computer networks. These initiatives were directed at nongovernmental organizations and several universities.

Ted Hope, of the Huracán project (see Chap. 3), describes Puliatti as a "driving force" behind the Internet in the region. The association between

Puliatti and Hope resulted in a productive working relationship. With Puliatti's support and contacts from the United Nations Development Programme, Hope traveled (often alongside NSRC's Randy Bush) to numerous Latin American countries to install low-cost equipment to connect the countries to computer networks (from UUCP to the Internet). This was accomplished in a dozen countries of the Latin American region.

A Sustainable Development Network

Parallel to Puliatti's work, the Sustainable Development Networking Programme (SDNP), another UNDP project, worked to encourage network access in the early 1990s. The concept of the SDNP was developed as a result of the UN Conference on Environment and Development celebrated in Rio de Janeiro in 1992. One of the outcomes of the conference was Agenda 21, an environmental and developmental action plan. Chapter 40 of Agenda 21 centered on the importance of information access as a social development mechanism. Said chapter began with the words: "In sustainable development, everyone is a user and provider of information considered in the broad sense. That includes data, information, appropriately packaged experience and knowledge" (UNDSD, 1992, p. 348). One of the suggested activities to reach this idea was the "establishment and strengthening of electronic networking capabilities" (UNDSD, 1992, p. 352). The UN Development Programme designed several initiatives to implement the agenda's established principals, one of which, Capacity 21, went specifically directed to development countries. In that context, connection to computer networks was seen as a means—or what Raúl Zambrano, technical advisor and Sustainable Development Networking Programme promoter in Central America called "an excuse" (interview with the author, December 14, 2017)—to achieve the projects' ultimate objectives, notably the flow of information as a central dynamic of development.

To implement the initiative, the Sustainable Development Networking Programme typically began with a feasibility study in each country. That way, ideal local counterparts were sought, who could be organizationally located in different types of entities (such as governments, universities, NGOs). The initiative's objective was to have these counterparts operate with sufficient autonomy based on each country's characteristics. Discussions usually initiated at a governmental level, in search of approval to operate (and eventually to negotiate with state telecommunications operators). Similarly, synergy was sought after with local counterparts sharing similar interests. The project offered some funding for local nodes, but a need persisted for finding a business model that permitted subsidizing once funds ran out. Typically, the project emphasized the need to develop two professional profiles in each local node: a coordinator with administrative duties and a technical assistant in charge of technological development.

After identifying counterparts came the installation of computer networking connections and the appropriate training to operate them autonomously. With regard to the first, the project sought to develop "appropriate technology," that is to say, "computer and networking technologies that [were] adequate for existing infrastructure and available human resources" (Daudpota & Zambrano, 1995, p. 1). Implementing this technology required negotiation with local telecommunications operators, always a challenging task. Although the project promoters were interested in providing Internet access, the costs of a dedicated connection seemed prohibitive. For that reason, in most Central American countries, it was decided to first provide UUCP access. Similarly, the need to work with Linux was emphasized, along with an array of free operating systems and open-source software. In the end, the project took a chance on creating capacities in the collaborators of each local node that could strengthen its scope and, in that way, reach the project's objectives. At least from 1993 to 1995, the project became affiliated with Internet Society courses about free software and participated in the INET conferences.

The Sustainable Development Networking Programme operated under a variation of this framework in all of the region's countries. Faithful to their principal of autonomy, these nodes acquired diverse configurations and had varied scope. In Costa Rica (1996) and Guatemala (1997), for example, the project joined existing initiatives at universities: the Observatory for Development at the University of Costa Rica, and Universidad del Valle in Guatemala, respectively. In Honduras (1994) and Nicaragua (1994), the project operated independently. Undoubtedly, the country of the region where the project acquired the greatest impact was Honduras. In this country, the node was installed at the UN offices. Zambrano turned to Ted Hope, from the Huracán project, for the equipment installation, which was carried out in August 1994. This node provided email access and connected to the UN Development Programme offices in New York twice a day to exchange messages. Prior to this link, some NGOs had email access through Nicarao and the two RACSAPAC nodes installed in Honduras (see Chap. 3). Thus, the Sustainable Development Networking Programme became the first local provider of email access.

Finding users for the project required a communications campaign directed primarily toward the country's nongovernmental organizations. In the words of Raquel Isaula, coordinator of the local SDNP:

> What we did was organize people into groups. We grouped [them] in discussion lists and information distribution lists and brought the first 25 modems, and we took them from office to office to connect them. If they didn't connect in a week, didn't send a message, we'd go get [the modem] and take it to another office. (Interview with the author, November 09, 2017)

By 1995, 64 local organizations were linked to the Honduran Sustainable Development Networking Programme (Afonso, 2002). As part of this initiative, the Honduran TLD was originally delegated to the Programme (which would become the focal point of a long-standing dispute in the country).

A regional perspective for the project's development was promoted from the start of the project's coordination. Those responsible for each regional node were called to biannual meetings where topics of common interest were discussed. Once the UN Development Programme's funds ran out, local projects had to find an economic model that would give them permanency and viability. In that sense, the region's Sustainable Development Networking Programme projects, notably in Guatemala, Honduras and Nicaragua, would have an intriguing trajectory that would transform them into access and Internet-service providers in the second half of the decade (a process that will be examined in more detail in Chap. 5).

A Front Divided by a Common Goal

Although they had practically identical objectives, the relationship between OAS and UN Development Programme initiatives was far from harmonious. Several people interviewed for this project characterized the relationship as "conflictive," "competitive," "rivalrous," and "antagonistic." (The dispute over TLD management and the Internet connection process in Honduras constitute typical expressions of this antagonism, discussed in the following chapter.) Ultimately, both projects were guided by relatively different approaches toward network connection. Just as the Huracán project or the Sustainable Development Networking Programme initiative illustrated, the UN Development Programme advocated for a teamwork philosophy with low-budget equipment that would allow a greater number of countries to connect rapidly and grow gradually in infrastructure. In that way, the use of dial-up connections and technologies like UUCP was prioritized in order to have quick access to networks like the Internet, since greater additional economic investment was not needed, and, simultaneously, the existing conditions in the Central American area were harnessed. It was, therefore, "appropriate technology" for each country (Daudpota & Zambrano, 1995, p. 1).

Nonetheless, the OAS-led project (which de Téramond and his collaborators supported) considered that these types of systems underestimated regional telecommunications infrastructure and limited the networks' long-term possibilities for growth. Typically, the Hemisphere-Wide Inter-University Scientific and Technological Information Network worked with high-performance equipment acquired from CISCO but at a high price. De Téramond exposed this difference of perspectives in 1994: "This type of [UUCP] connection, useful when there is no existent telecommunications infrastructure in a country, was discarded in our project from the beginning in favor of high-capacity technology and dedicated connections, given the great number of predicted users and the type of use that our researchers needed from the system" (1994, p. 65).

In that sense, computer networks were carried out through various political projects in Latin America. Although they sought common goals through these networks, such as integration, development and scientific progress, the ways of achieving these objectives—and, therefore, the proposed technologies—were not always shared. In practice, all of these initiatives, political visions, and ambitions came together in local Internet connection projects that acquired nuances and specific configurations.

Concluding Remarks

This chapter took into consideration the ways in which institutional conditions were established for Internet access in Central America. With that in mind, the initiatives of three organizations that developed networking initiatives over the course of the 1990s were discussed: the National Science Foundation, the Organization of American States, and the United Nations Development Programme.

The role of organizations like NSF, OAS, and UNDP can be thought of as an instance of what van der Vleuten (2006), based on Hughes' (2004) work, calls "institutional system builders." As such, these organizations participated actively in the configuration of the Internet in Central America. Their transnational action capacity was established in three dimensions:

(a) Coordination: the ability to identify relevant actors in various countries, create work networks between them based on common interests, form institutional alliances and configurate spaces or transnational "contact zones" (Sandoval García, 2009) (such as conferences, seminars, workshops, meetings, etc.) where potential allies were found, solutions were thought of, and agreements were reached among actors from various countries.

(b) Funding: the ability to obtain sufficient funds for distribution in different parts of the American continent according to specific needs. These funds were vital for resolving some of the primary obstacles in the crusade for computer network implementation in the Latin American region: buying basic equipment, strengthening existing infrastructure, and obtaining indispensable knowledge to establish the connections, as well as supporting them.

(c) Negotiation: political power that would allow for the creation of dispositions and recommendations that would give network connections sufficient institutional relevance, the acquisition of required permits for installing equipment in the country with the endorsement of state telecommunications monopolies, and the improvement of equipment tariffs to make it more accessible in the Latin American region.

Van der Vleuten (2006) argues that, as institutional systems builders, the work of these types of organizations is characterized by ideological, sociotechnical and controversial aspects. All of these dimensions intermingled in the proposed initiatives for facilitating Latin American Internet access. NSF and OAS built an alliance that emphasized the potential of computer networks for advancing science in Latin America and promoted the connection through high-performance equipment to improve its growth in the mid- and long term. This perspective conflicted with the UN projects centered on the promise of networks like the Internet for social change and development, which emphasized the necessity of adjusting the acquisition of equipment to the reduced economic possibilities of regions like Central America. It was then, essentially, about different political understandings of the Internet as a sociotechnical project.

On the basis of this transnational action capacity mobilized by various international organizations, the Internet found relatively fertile ground in the Central American region. The following chapter offers a detailed discussion of how promoters and collaborators in each country put their projects and initiatives into practice to finally achieve connection.

NOTE

1. Abbate notes that "in the late 1980s the US portion of BITNET gave up the RJE system in favor of TCP/IP, and a version of UUCP was developed to run over TCP/IP. In 1986 the BITNET and UUCP organizations also agreed to adopt the Internet's Domain Name System" (p. 205).

REFERENCES

Abbate, J. (2000). *Inventing the internet*. Cambridge, MA: MIT Press.

Abbate, J. (2010). Privatizing the internet: Competing visions and chaotic events, 1987–1995. *IEEE Annals of the History of Computing, 32*(1), 10–22.

Afonso, C. (2002). *SDNP evaluation report: Colombia, Honduras and Nicaragua*. New York: UNDP.

Daudpota, I., & Zambrano, R. (1995). *The sustainable development networking programme: Concept and implementation*. Paper presented at the INET '95 conference, Honolulu, Hawaii.

de Téramond, G. (1994). Interconexión de Costa Rica a las grandes redes de investigación Bitnet e Internet. In *Ideario de la ciencia y la tecnología: hacia el nuevo milenio* (pp. 61–86). San José, Costa Rica: Ministerio de Ciencia y Tecnología.

de Téramond, G., & Brenes, A. (1994). *Establecimiento y consolidación de los proyectos Internet en Centro América y el Caribe bajo el marco del proyecto RedHUCyT*. San José, Costa Rica: CRNet.

Gillespie, T. (2007). *Wired shut: Copyright and the shape of digital culture*. Cambridge, MA: MIT Press.

Goldstein, S. N. (1991). *Federal and international directions in computer networking*. Paper presented at US Sprint, Reston/Herndon, VA.

Goldstein, S. N. (1995). *Future prospects for NSF's international connections program activities*. Paper presented at the INET '95 conference, Honolulu, Hawaii.

Hafner, K., & Lyon, M. (1998). *Where wizards stay up late: The origins of the internet*. New York: Touchstone.

Hahn, S. (1995). *The Organization of American States hemisphere-wide networking initiative.* Washington, DC: Organization of American States.

Hahn, S. (1996). Networking in Latin America and the Caribbean: Recent advances of the OAS/RedHUCyT project. *IGLU, 10*(April), 53–61.

Hahn, S. (1999). Case studies on developments of the internet in Latin America: Unexpected results. *Bulletin of the American Society for Information Science, 25*(5), 15–17.

Hahn, S. (2000, July 18–21). *Case studies on development of the internet in Latin America and the Caribbean.* Paper presented at the Internet Society (INET) conference, Yokohama, Japan.

Hughes, T. P. (2004). *Human-built world: How to think about technology and culture.* Chicago: University of Chicago Press.

Lehtisalo, K. (2005). *The history of NORDUnet.* Kastrup, Denmark: NORDUnet.

Organization of American States. (1994). *Plan de acción de la Primera Cumbre de las Américas.* Washington: Organization of American States.

Press, L. (1996). Seeding networks: The federal role. *Communications of the ACM, 39*(October), 10–18.

Sandoval García, C. (2009). Zonas de contacto entre las ciencias sociales. In M. Baltodano, E. Cook, & R. Mooney (Eds.), *Género y religión: Sospechas y aportes para la reflexión* (pp. 177–191). San José, Costa Rica: Editorial SEBILA.

Siles, I. (2008). *Por un sueño en.red.ado. Una historia de Internet en Costa Rica (1990–2005).* San José, Costa Rica: Editorial de la Universidad de Costa Rica.

Siles, I. (2017). 25 years of the internet in Central America: An interview with Guy de Téramond. *Internet Histories, 1*(4), 349–358.

SNDP. (n.d.). *Using information technology for sustainable development.* Reporte inédito.

Soriano, J. (2007). Steve Goldstein de NSF: Network training workshops. *Interred,* from https://interred.wordpress.com/2007/01/23/steve-goldstein-de-nsf-network-training-workshops/

UNDSD. (1992). *Agenda 21: United Nations conference on environment & development.* New York: United Nations Division for Sustainable Development.

van der Vleuten, E. (2006). Understanding network societies: Two decades of large technical system studies. In E. van der Vleuten & A. Kaijser (Eds.), *Networking Europe: Transnational infrastructures and the shaping of Europe, 1850–2000* (pp. 279–314). Sagamore Beach, MA: Science History Publications/USA.

Yeung, L., & Puliatti, E. (1992). UNDP. *Internet Society News, 1*(2), 22.

CHAPTER 5

A Central American Internet

Abstract This chapter analyzes in detail how each country of Central America connected to the Internet. The discussion follows two parallel processes. On the one hand, the chapter explains the singularities of local connection projects in each country. It analyzes the motivations and challenges involved in local networking initiatives. On the other hand, the chapter also focuses on the transnational flows of people, knowledge, and technologies that cut across the isthmus to make local projects possible. More than departing from a transnational approach, the discussion of each country individually demonstrates how transnational flows and exchanges between countries materialized in specific ways at the local level. The chapter makes visible the singular approach to integration that becomes clear when regional networking initiatives are examined through a transnational lens.

Keywords Central America • Computer networks • Internet • Transnational history

To establish connection in Central American countries, the "momentum" generated by the institutional projects described in the previous chapter was used (Hughes, 1994). In the working framework provided by that array of initiatives, the region's countries connected to the Internet through dedicated links from 1993 to 1996. This chapter analyzes how

© The Author(s) 2020   79
I. Siles, *A Transnational History of the Internet in Central America,*
1985–2000, Palgrave Macmillan Transnational History Series,
https://doi.org/10.1007/978-3-030-48947-2_5

those connection efforts emerged and were put to the test. The discussion is organized chronologically to reflect the order in which the Internet link was established in each country of the region.

Central American interconnection *to* and *through* the Internet was characterized by a combination of three great sources of action: local networking initiatives, international organizations, and "network entrepreneurs" (Burt, 2000). To fully understand this interconnection process, it is essential to consider how people, knowledge, and technologies were mobilized across national borders. In that sense, the analysis of each country is not intended to depart from a transnational approach. On the contrary, the purpose of each analysis is to demonstrate the ways in which these flows and transnational exchanges acquired specific configurations in each of the region's countries.

A perspective regarding integration becomes clear when discussing these common connection processes. The aim was to create sociotechnical networks where people, knowledge, and technologies could circulate so that computer networks could be viable and self-sufficient, and thus strengthen the region's development. In that sense, this chapter tells the original and inedited story of the Internet in Central America and the foundations that were laid to bring about regional development.

THE COSTA RICAN CATALYST

After the pioneering experience with BITNET, described in Chap. 3, the task force led by de Téramond at the University of Costa Rica focused on achieving Internet connection, which was "completely unknown territory for those in charge of the project" (de Téramond, 1994a, p. 72). In his summary of the Costa Rican interconnection project published in 1993 by *Internet Society News*, de Téramond reiterated the vision of computer networking that guided his work since the days of BITNET. In his opinion, the Internet represented a tool for overcoming academic isolation and strengthening the country's development:

> Without access to the world scientific potential, scientists in the country will remain isolated, frustrated and unproductive. This project represents an opportunity to learn and implement scientific communication tools as well as giving a major impact to the whole education system and country economy. (de Téramond, 1993)

The decision to connect to the Internet marked a clear detachment from the work of Costa Rican state telecommunications companies, which had invested money and efforts into the development of a X.25 public data network (see Chap. 3). This technical divergence reflected distinct political visions, which Abbate summarize with precision:

> The [X.25 y TCP/IP] protocols had not been designed to work together, and combining them would needlessly duplicate many functions; they were clearly meant to be alternative approaches to building networks… Though each provided a system for networking computers, they embodied different assumptions about the technical, economic, and social environment for networking. The tension between these two visions manifested itself as a battle over standards. (Abbate, 2000, p. 155)

The X.25 network was mainly criticized for its limitations in interconnecting with other computer networks, by way of the X.75 protocol. While Internet promoters could interconnect to the highest number of networks possible, regardless of how different they were, telephone companies tried to achieve interconnection by making each existing network adjust to X.25 procedures (Abbate, 2000, pp. 152–167; Gillespie, 2006, p. 432). Although this would change very quickly, the local Costa Rican telecommunications operator showed little interest in networks that to them seemed to be an academic project with an unclear outcome (or even a mere entertainment for academics).

To connect to the Internet from Costa Rica, working methods similar to those learned during the link to BITNET were adopted (Siles, 2008). The "heterogenous engineering" developed by de Téramond can be summarized as three processes (Law, 1987). First, sufficient funds were needed to acquire routing equipment and to pay for a greater capacity satellite link to Homestead, Florida. The lack of funding to connect to a relatively unknown technology in the country had been one of the primary challenges faced during previous initiatives. This was resolved through a donation from the International Development Association (IDA) of $150,000. Obtaining these funds required, once again, the formation of a network of transnational connections. Larry Landweber, professor of computing at the University of Wisconsin-Madison, mobilized contacts at his university, CISCO and the International Development Association, which was crucial to developing the funding project (de Téramond, Brenes, Espinoza, & Bonilla, 1997).

Second, the project needed scalability in the immediate future as well as political and academic legitimacy. Resolving these issues was not a simple task. The myriad existing networks did not make it easy to anticipate the most viable technological path in the mid- and long term. As a solution, de Téramond encouraged creating a national network that would include various institutions associated with academic and scientific research, interconnected via Internet: the Costa Rican National Research Network (CRNet). De Téramond (1993) described CRNet as "digital network that will provide high speed local interconnectivity between scientists at universities, research laboratories, industries with a technological component and other national institutions." Hence, the fund request to the International Development Agency was presented by the University of Costa Rica, the Costa Rica Institute of Technology (ITCR), the Tropical Agricultural Research and Higher Education Center (CATIE), and the University of Wisconsin-Madison. The creation of CRNet formed part of this funding request. Finally, following difficult negotiations within the University of Costa Rica, de Téramond was able to consolidate his task force: he was given an office at the Computer Center, three collaborators were added to the group, and one of the engineers (Jorge Wing Ching) was sent to the University of Wisconsin-Madison to familiarize himself with Internet technologies.

In spite of these advancements, the arrival of Hurricane Andrew in Florida delayed connection plans, as it affected the receptor antennas in Homestead. Finally, on January 26, 1993, twelves nodes installed at the Computer Center, the School of Geology, and the School of Physics at the University of Costa Rica connected to the Internet. In addition to a router loaned by the University of Wisconsin-Madison, some UCR professors made their personal computers available. Using open source software, these loaned computers served as routers to connect to Internet. Abel Brenes remembers when the connection was established: "We were in the room where [the IBM computer] and the table were, and Guy [Téramond] was in his office, which was two cubicles away. [Suddenly] Guy comes out running, shouting out: "A packet! A packet!" At that moment we started seeing traffic, and the connection was established" (Interview with the author, May 23, 2006).

The next day, de Téramond communicated the news in an email directed to a group of collaborators. The group of email recipients, which included Steve Goldstein (NSF), Larry Landweber (University of Wisconsin-Madison) and Saúl Hahn (OAS), conveys the transnational network of people and organizations behind Costa Rica's connection. The message said:

> A 64 kbps-channel was established yesterday between Costa Rica and the NSF/Sprint/Panamsat POP in Homestead, interconnecting CRNet (AS 2146) to the Internet. The CRNet routes were propagated to Alternet, SprintLink, Suranet, EBONE (parts of Europe), PSI, parts of Japan, and Ecuador. The routing associated with NSFNet will not be active until Friday morning, when NSFNet updates its routes.

Network access was officially inaugurated at the end of April 1993, through a ceremony with university officials. The ceremony was the closing activity for a training course about routing system operations and Internet protocols, sponsored by the University of Costa Rica and the Organization of American States. The course had 44 participants, 10 of whom came from other countries of the region (Guatemala, Honduras, Nicaragua and Panama). The seminar included the following topics: modems and means of Internet transmission; TCP/IP protocols; hardware and routers; routing protocols; nameservers; and applications available via the Internet.

In the months following the connection, de Téramond's team worked to take CRNet from concept to practice. The CISCO routers intended for the connection were delivered by customs in March 1993. Abel Brenes, the person in charge of routing, took on the responsibility of configuring them. The work procedure was given the name "backbone on a table" since, as the name implies, de Téramond and his collaborators carried out the planning and configuration of the equipment by first installing it on a table that depicted a map of the connection. Fig. 5.1 shows those responsible for the project with the connection equipment set up on a work table.

In April 1993, only three months after the first connection in the country, the project's collaborators established a link between the Computer Science Research Center at the Costa Rica Institute of Technology, the National Distance Education University (UNED) and the University of Costa Rica, with the first IP network in Central America: CRNet.

The Costa Rican Research Network (CRNet) grew rapidly. De Téramond described the project's advancements from 1993 to 1994 enthusiastically, in an edition of the *Internet Society News*:

Fig. 5.1 *"Backbone on a table"*: Planning CRNet's connection structure. (In this photograph (left to right): Abel Brenes, Ana Lucía Chavarría, Guy de Téramond and Mario Guerra. Source: Guy de Téramond Peralta. Used with permission)

> Undoubtedly 1993 represents a landmark in the history of computer communications in Costa Rica with the interconnection to the Internet, and the creation and consolidation of CRNet [...] The number of connected nodes to CRNet increased from 12 in January 1993 to 250 at the end of the year. [...] the number of nodes will double during the first weeks of 1994. A rapid expansion into the commercial sector and doubling the speed of all the links is expected. (de Téramond, 1994b)

With the support of the OAS Hemisphere-Wide Inter-University Scientific and Technological Information Network project, in May 1994 new equipment was acquired, making it possible for the network to expand and offer connection to more organizations. That year, CRNet connected 30 institutions. To finance the project, members paid a quarterly fee based on the bandwidth requested. The group in charge of the connection offered training sessions about Internet operations. The most used applications included email, Gopher, Archie, and WAIS. Thus, this was mostly about file exchange and conferring about documents through text-based interfaces.

The enthusiasm that this Internet connection generated in the Costa Rican academic community was reflected in a compilation of essays published by the Ministry of Science and Technology in 1994 (MICIT, 1994). For instance, in this document, Max Cerdas, Chief Information Officer at Costa Rica's National Council for Scientific and Technological Investigation, and who had collaborated on the BITNET connection project, catalogued the network's potential as "unimaginable" and its resources as "unlimited" (Cerdas, 1994, p. 19). The director of the Omar Dengo Foundation added: "Telematics has made it possible to overcome geographically-imposed limitations" (Fonseca, 1994, p. 134). The first book published in the country to popularize Internet use employed similar discourse. Its author, who had coordinated connection from the Costa Rica Institute of Technology to the Internet, compared Internet access to a "mine," a "reservoir," and a "New World" (Bogarín, 1994a, p. 8). The book's cover depicted three caravels approaching a verdant shore (as opposed to the arid lands that were their point of departure).

At the same time, these texts suggested that Costa Rica had to assume a leadership position in the discovery of this new world made possible by the network. Thus, the compilation of essays responded directly to an invitation from the Minister of Science and Technology at that time (Orlando Morales), to transform the country into "information capital." In his contribution to the MICIT book published in 1994, Rodrigo Bogarín (1994b), of the Costa Rica Institute of Technology, stated: "achieving Internet connection is not an end in itself, but a starting point [...] to convert our country into an Information Center of regional importance" (p. 94), an invitation to consider Costa Rica the center of the region's telecommunications sector. This ambition can be read as a political concern in the context of the era's new regionalism (see Chap. 2).

The case of Costa Rica served as a "catalyst" for the region's Internet development in two significant ways. First, it allowed for the consolidation of a local task force with knowledge that, with support from the Hemisphere-Wide Inter-University Scientific and Technological Information Network project, would have a multiplier effect in the region. The establishment of an academic network connected to the Internet (CRNet) served as a model adopted and adapted by other countries in the region, with aid from the Organization of American States. In this process, de Téramond served as a crucial "network entrepreneur" in Central

America who would act as intermediary among the actors and regional groups interested in Internet connection. Additionally, the engineers of the Costa Rican project became the OAS's operating branch in Central America.[1] Finally, the project in Costa Rica installed a direct access node to the Internet through which Nicaragua and Panama would connect several months later. The following sections explain this process in more detail.

NICARAGUA AND THE FIRST "PURE LINK" TO THE INTERNET

One of the most significant challenges faced by local projects was negotiating with each country's government and state telecommunications company, a process typically complicated by the lack of awareness of the Internet as a technology, by its then-academic nature, and by the investments dedicated to other projects, notably the X.25 network (Abbate, 2000; Siles, 2012). In general, those involved in each country's projects explain the difficulties with a common phrase: the Internet was seen as a threat to telecommunications companies' commercial interests. Enzo Puliatti, of the UN Development Programme, describes it thusly: "It was like being missionaries, evangelists who knocked on people's doors trying to convince them" (interview with the author, November 02, 2017). Furthermore, while these organizations dealt with the mystery surrounding the Internet, they faced another threat as well: privatization. From 1995 to 1996, every country in the region (except Costa Rica) approved legal reforms aimed at privatizing the telecommunications market in some way. (This topic will be examined in more detail in Chap. 6).[2]

The case of Nicaragua, the next country in the region to connect, is illustrative in this sense. Once the link through UUNET had been established (described in the previous chapter), the following objective was to connect directly to the Internet. Hopmann recruited Teresa Ortega, an engineering student at the National University of Engineering, to coordinate this initiative. Because of the Nicaraguan Revolution, in which many men were involved, the majority of engineering university students were women. Ortega recalls: "I got [to the university] and there were [approximately] 100 in a room, when I was 18 in [1983]. And they started calling all of the guys to fight, to the north, to the mountains. In the end, there were only 9 of us left, and we were all women" (interview with the author, July 30, 2017).

By May 1989, the National University of Engineering had been granted administration of the top-level domain (.ni), thanks to Hopmann's efforts in the UUNET connection process.[3] It was the first country in Central America to obtain said administration (and the fifth in Latin America). At the end of 1991, after participating in the conferences that had taken place in the previous months, the work teams at the University of Costa Rica and the National University of Engineering had elaborated a preliminary document where a plan to achieve the connection was developed. The idea was to connect Nicaragua through the microwave link toward Costa Rica and utilize the Costa Rican international channel to connect to the Internet. This represented a relatively less onerous option than establishing a satellite link from Nicaragua. In his "Description of the Research Network Initiative in Costa Rica," published by *Internet Society News* in 1993—the year prior to the execution of this plan—de Téramond made his expectations and ambitions about regional interconnection clear: "It is expected that this connection will be the starting point of a Central American internet backbone" (de Téramond, 1993). His 1994 report to the Organization of American States expressed this idea practically with the same words: the connection plan between the two countries was seen as "the first step [toward] a Central American backbone" (de Téramond & Brenes, 1994, p. 87).

In 1992, the OAS RedHUCyT project approved the budget to finance this plan and a discount was given on routing equipment through SINSA, a company representing CISCO in Costa Rica. Nonetheless, the negotiations with the Nicaraguan government and the local telecommunications company (TELCOR) were significantly more complicated. In the words of Yadira Mena, computer engineering student at the National University of Engineering (UNI) who assumed coordination of the connection project while Teresa Ortega prepared to study abroad, "getting the government to support us was a daunting task […] having instant communication threatened traditional snail mail as well as both domestic and international calls" (interview with the author, July 14, 2017). Similarly, Hopmann defined these relations as "[nine months of] taking two steps forward and one step back with TELCOR, UNI and MEDE [Ministry of Economics and Development], regarding who was in charge of administrating assets" (1998, par. 3). Numerous meetings were needed to clarify the legal status under which the Internet would operate, and multiple efforts were needed to convince local authorities.

In March 1993, Hahn and de Téramond traveled to Managua to support the conversations with TELCOR, the Nicaraguan government and the National Engineering University. The project did not receive approval from TELCOR until the Internet Society's INET conference in August 1993, held in San Francisco, California. Mena recalls:

> During a break, we managed to discuss the topic [with the TELCOR representative]. I already had the document ready, a white sheet, no letterhead, only the terms and [the clause that indicated] that they agreed with going ahead with the project. We got the signature right there. (Interview with the author, July 14, 2017)

The fact that said memorandum of agreement could only be achieved outside of Nicaraguan territory is not a coincidence. The intermediation of OAS and NSF representatives in the meeting helped obtain this agreement. The memorandum of agreement was signed by Mariano Buitrago (director of the TELCOR computer science division), de Téramond, Hahn, and Teresa Ortega.

After this agreement, according to Hopmann (1998), the necessary equipment for the connection was held for five months by Nicaraguan customs. To establish the connection, three Costa Rican engineers (Abel Brenes, Roger Brenes y Mario Guerra) went on a five-day trip to Nicaragua with funds from the OAS RedHUCyT project. Roger Brenes describes the typical process that characterized this work:

> When we arrived, I took charge of connecting the institution's modems with the central unit, which could vary depending on the country, but it was almost always at a telephone company [...] Once we made connections at the modem level, we connected routers, and Abel [Brenes] successfully configured the equipment from a remote location, and Mario [Guerra] took responsibility for services like email and the nameserver. (Interview with the author, February 08, 2006)

The link from Nicaragua to the Internet via Costa Rica's international infrastructure was established in February 1994. Due to the nature of this technological solution as the first of its kind in the Latin American region, Arce and Hopmann (2002) did not hesitate to categorize it as "the first pure [Internet] link in history" (p. 24). The connection was celebrated by an official inauguration ceremony, described by Yadira Mena in the following way: "The media were invited, [political figures] were invited. A tour was given [to show] all the students the novelty and pride that was the first university with Internet in Nicaragua" (interview with the author, July 14, 2017). Six months later, in October 1994, Abel Brenes and Roger Brenes returned to Nicaragua to install additional routers that made it possible for two more universities to connect to what was named the Nicaraguan Academic Internet Network (RAIN).

Despite Nicaragua's early advances in computer networking—access nodes to UUCP and UUNET had been available since before the start of the decade—Internet development in the second half of the decade was relatively slow compared to other countries of the isthmus. The reasons for this can be found in the same factors that made the connection to the Internet so difficult: a lack of governmental support and of awareness at an institutional level about the value of the network to enable activities conducive to academic work and to foster Nicaragua's productive sector in the long term. As will be discussed in Chap. 6, the rise of the Internet also coincided with the opening of the telecommunications sector to private participation in the country, a process that turned out to be more complicated than expected.

ANOTHER RUNG ON THE TRANSNATIONAL LADDER: THE INTERNET IN PANAMA

Internet connection for the remaining Central American countries involved similar negotiations with governments and state telecommunications operators. An initial discussion about an eventual link to the Internet from Panama took place at the 1992 INET conference between the OAS RedHUCyT project and the Panamanian engineers in charge of the BITNET project. Víctor López, one of the project's leaders, recalls:

> Saúl Hahn came up to me and introduced himself. He [asked] if there was any interest in Panama joining the OAS RedHUCyT project, which I immediately replied yes to. We made the decision to do it right then and there, although later I went to the chancellor [at the Technological University of Panama], who was in charge [of the project] to tell him: "They offered us this. They're going to provide the equipment, funding [and] support." (Interview with the author, July 07, 2017)

The Internet connection from Panama extended the working relationship with Costa Rica that had been instrumental in establishing the node to BITNET at the Technological University of Panama (in 1992) (described in Chap. 3). This resulted in a constant flow of people and knowledge between the two countries. In April 1993, as the institutional terrain necessary for the Panama connection was being prepared, several engineers traveled to the University of Costa Rica to participate in Costa Rica's Internet inauguration and the training course in Internet protocol operations. The event also provided the opportunity to coordinate the details of the connection in Panama. Julio Lezcano, in charge of technical aspects of this project, recalls: "That week, we sat down to talk with Guy [de Téramond], Saúl [Hahn], Roger Brenes [SINSA] [...] to take a look at the equipment that we needed. Practically the only thing left to do was schedule when to make the transition [to the Internet]" (interview with the author, July 07, 2017). Three months later, in July 1993, Hahn and de Téramond traveled to Panama to settle on the details: the plan was to substitute the BITNET link for a connection between the Costa Rican Research Network (CRNet) and its Panamanian equivalent (by the name of PANNet). In addition to the Technological University of Panama, the project's leader and recipient of the TLD administration, both the University of Panama and the Santa María la Antigua University joined the initiative.

Following negotiations with the state telecommunications company (INTEL) and the acquisition of the necessary equipment—thanks to the funds provided by the OAS—Abel Brenes, Roger Brenes, and Mario Guerra traveled from Costa Rica to Panama for five days to participate in the connection, which was established in June 1994. During the first year, Panama was connected using Costa Rica's international channel through the analog microwave between both countries, as was Nicaragua. The OAS RedHUCyT project financed the cost of the microwave link for a year and increased the bandwidth of Costa Rica's satellite channel to facilitate the Panamanian connection. In addition to configuring the equipment, the visit from Costa Ricans engineers to Panama included a training course about "routing systems and Internet protocols" (de Téramond & Brenes, 1994). The process concluded with the usual inauguration of the Internet before the media, university staff, and political figures.

During the following months, work focused on spreading the technology among the connected universities and throughout the rest of the country. Inmaculada de Castillo, professor in computer science at the Technological University of Panama and administrative director of the project, narrates:

[The priority was] to provide continuous training for the people who support[ed] the networks and to also create an entire awareness raising process regarding the network's importance. In Panama, we traveled almost border-to-border to make the network's value known and to promote the interconnection of academic institutions. (Interview with the author, July 07, 2017)

The 1994–1995 Internet development process coincided with the early boom of the World Wide Web in the region. The Web facilitated graphic interfaces to access information and, due to the rupture that this represented for text-dominated environments, its emergence marked "a new paradigm" for how the network was used (de Téramond, cited in Siles, 2017, p. 354). During 1995, the CRNet team returned to Nicaragua and Panama to offer an "advanced workshop in the administration of TCP/IP and Internet environments" (de Téramond & Brenes, 1995). On those occasions, the visit included an additional activity: configuring the country's first websites. Regarding the efforts to promote the use of the network and Web, Hahn recalls an initiative undertaken at the Technological University of Panama, which he witnessed while visiting Panama some weeks after establishing the connection: the halls of the university were filled with images of Jupiter downloaded from the Web, which had been published by NASA following one of its recent observation missions to that planet.

The Internet and the Conflictive Top-Level Domain in Honduras

The following country to connect was Honduras. A starting point to establish the connection was the formation of the "Inter-Tegus Group" in 1993 at the National Autonomous University of Honduras (UNAH). The group united 40 individuals interested in the Internet connection process. At the center of the efforts to form and coordinate the group was Gustavo Pérez, professor of physics at this university. Apart from a long personal history in computer use, two concrete experiences marked Pérez's relationship with computer networks: first, he had been a Huracán user (as was indicated in Chap. 3) and, second, Pérez had spent 1992 on a research stay at the European Organization for Nuclear Research (CERN), in Switzerland.

The Inter-Tegus proposal went beyond merely establishing the Internet connection. Its original goal, explains Pérez, was "to utilize the university's special autonomous status to place a free communications area in the campus, connect to the satellites, and offer to local businesses a place where they could connect and bring their equipment tax-free, and offer services" (interview with the author, December 04, 2017). The group met approximately twice a month to discuss aspects related to this project.

One of the main results of this group's formation was the formulation of a concrete goal to establish the Internet link. To bring this to fruition, Pérez turned to his professional contacts. Pérez knew Saúl Hahn since the Organization of American States had financed a research project that included establishing a laboratory in Honduras. Pérez adds: "When the Internet project began, I called Hahn and he told me that Guy de Téramond, who I also knew as a physicist, had established the connection in Costa Rica" (interview with the author, December 04, 2017). Additionally, during a personal trip that Pérez took to Miami, he took the opportunity to contact Brien Morgan, PanAmSat engineer, with whom he met to discuss the requirements for connecting to Honduras with the Homestead Internet node, and to invite him to the country.

Inter-Tegus also sought aid from the Honduran National Minister of Science and Technology, Humberto Consenza. On Consenza's recommendation, the group turned to the National Council for Science and Technology (COHCIT), formed that same year, to inquire about funding possibilities. As this dealt with a governmental agency—and in light of what some Inter-Tegus members considered a lack of support and understanding on part of the National Autonomous University of Honduras authorities—this council gradually came to oversee and manage the project. In that web of conversations, the National Autonomous University of Honduras extended a formal invitation to de Téramond, Hahn and Morgan, who traveled in September 1993 to meet with Inter-Tegus members and other interested institutions, and thus discuss the possibilities of implementing the plan. At Consenza's invitation, de Téramond stayed in Honduras a few days more, honing the details of the country's proposed Internet connection.

As a result of these numerous meetings, the final proposal was ready in October 1993. The plan proposed connecting to the Internet to universities, laboratories, related industries, and other national institutions (de Téramond & Brenes, 1994, p. 99). As originally planned, the Honduran network HONDUNet would connect by satellite link to the antenna in Homestead, Florida. To finance the project, the OAS RedHUCyT project approved a preliminary budget of $40,000. As a second stage, connection to San Pedro Sula and La Ceiba, two cities in the northern part of the country, was proposed. The equipment acquired for the project was sent to Honduras at the start of 1995 and, in April of that year, de Téramond returned to Honduras accompanied by Roger Brenes to coordinate the connection details.

However, in the following weeks, the HONDUNet project was suspended due to a controversy regarding the country's top-level domain administration.[4] As was previously explained (Chap. 3), the TLD had originally been delegated to the Sustainable Development Network, after its role in the installation of the UUCP node in 1994. Those responsible for the HONDUNet project requested a change in administration on several grounds: the academic network included a range of academic and governmental institutions and, therefore, better represented those interested in managing the TLD; the TLD could not be bestowed upon the UNDP because it was not a local institution; and, the close working relationship between the Sustainable Development Network and HONDUTEL threatened compliance with the rules specified for TLD management. As for the Sustainable Development Network, it claimed not to have violated any Internet Assigned Numbers Authority rule and demanded that the original domain assignment be respected. Due to this difference, the project came to a deadlock.

After the impasse generated by this controversy, the connection to the Internet occurred in a fragmented manner. The Honduran National Council for Science and Technology (COHCIT) lent continuity to the HONDUNet project and, therefore, maintained affiliation with the OAS RedHUCyT work group. To resolve the top-level domain controversy, the council chose to reassess its strategy: it requested another domain (hondunet.net) to develop the project. With that issue resolved, in May 1995, Abel Brenes, Roger Brenes, de Téramond, Hahn, and Mario Guerra traveled to Honduras for four days to participate in the COHCIT and the National Autonomous University of Honduras connection to the Internet, as well as to offer training to local counterparts. Several situations complicated the connection process. Specifically, the work group found the HONDUTEL equipment to be relatively old and lacking in the conditions to connect to the necessary routing device. De Téramond explains the way in which this difficulty was resolved:

> Roger Brenes spent all night testing a cable, all possible connections, and welding each one of them. The number of possibilities is a factorial number. Roger spent all night trying each possibility. [...] In this way, we established

the connection to the Internet in Honduras, just before the inaugural event. My collaborators were pale for not having slept all night. (Interview with the author, June 23, 2017)

On June 6, 1995, the day that the Internet connection was established, Hahn communicated the news in an email: "It is our pleasure to inform that as of today, Honduras has been connected to the Internet network through a satellite link to Homestead, Florida. [...] The project [...] is the result of the participation of numerous organizations and individuals who together made this major step possible." In effect, this email reflects the complex web of transnational actors involved in the process. Hahn recognized the participation of Honduran, Costa Rican, and international actors as well as the Honduran National Council for Science and Technology, the National Autonomous University of Honduras, HONDUTEL, NSF, Sprint, the Costa Rican Research Network, the University of Costa Rica, SINSA, PanAmSat, and the Organization of American States.

Gustavo Pérez, who during 1994 had spent another research stay at the European Organization for Nuclear Research and who had returned to Honduras the following year, remembers the details of the Internet link's configuration at the National Autonomous University of Honduras campus: "I was able to buy several SUN stations. In some SUN stations we installed the tools necessary to connect [the] Physics [department] [and] Engineering; and the main server [was installed] at the COHCIT [Honduran National Council for Science and Technology]" (interview with the author, December 4, 2017). Over the coming months, other actors (including HONDUTEL) developed their own Internet connection initiatives. Thus, by 1996 there were various Internet-service providers: HONDUNet, HONDUTEL, the Sustainable Development Network itself, and IBM, via its GBNet network. In this way, the Internet's early development in Honduras was more fragmented than in the rest of the region's countries.

THE STRENGTH OF NETWORKS: THE INTERNET AS AN INTER-INSTITUTIONAL PROJECT IN GUATEMALA

The need to make concerted efforts among numerous actors, establish alliances to legitimize the project, and share connection costs functioned as a common driving force in many of the region's countries. This need was particularly felt in Guatemala. The possibilities offered by UUCP were rapidly exhausted and the connection of neighboring countries put pressure on Guatemala.

In 1991, the Guatemalan National Council of Science and Technology (CONCYT) created a Computer Science and Information Commission, made up of representatives from the then five Guatemalan universities, a professional association, and the Guatemalan Telecommunications Enterprise (GUATEL), an entity under state monopoly. The Commission quickly became convinced of the need for a project that allowed the interconnection of Guatemalan institutions and organizations with other countries via the Internet. The director of the commission was Luis Furlán (see Chap. 3). As in the case of UUCP, Furlán would assume an important role in the Internet link process and the implementation of the academic network. Francisco ("Frankie") Arzú, technical collaborator to the project at the Universidad del Valle, describes him as the "anchor point" between the various actors interested in the connection (interview with the author, November 08, 2017).

During Saúl Hahn's visit to Guatemala at the start of 1993, the commission put forward the MayaNet proposal, the same name given to its academic network. Originally, Hahn accepted financing the plan using the same solution implemented in Nicaragua and Panama, that is to say, through a microwave link toward Costa Rica, which would provide the satellite link to the Internet. However, according to Grete Pasch, who represented the Francisco Marroquín University in the commission, "the need for a satellite link to serve Guatemala" was "insisted on" (Pasch & Valdés, 1997, p. 7).

In spite of the initial impetus that characterized MayaNet's creation, Internet connection in Guatemala took several more years to come about. During a trip to Guatemala to participate in the National Computer Science Conference in March 1993, de Téramond and Hahn met with interested actors, but they did not settle on a connection plan because they identified a specific obstacle, which was stated in the visit's report: "it [was] not possible to involve [...] GUATEL" (de Téramond & Brenes, 1994, p. 13). Furlán also attributes this significant setback to the difficulties in working with the Guatemalan telecommunications company:

> GuaTel didn't allow any form of electronic communication to take place without its intercession, which slowed everything down for various years. Finally in [February] 1995, after a lot of lobbying to the National Congress,

the president of the nation, and his cabinet, an agreement was established and signed between CONCYT and GUATEL. (Furlán, 2007, par. 10–11)

The context for arriving at the agreement was complicated. At the start of the decade, an initial attempt to privatize telecommunications had been made, but it had failed because of the lack of aid from key sectors such as unions (Bull, 2005). (As discussed in Chap. 6, the following attempt would be more successful and would result in the signing of a liberal telecommunications law in 1996). Arriving at this preliminary agreement required numerous conversations with GUATEL authorities to resolve doubts about the network's operation and involved intervention from the country's government itself. To break the impasse, Magaly Morales, coordinator of Science and Technology at the National Council for Science and Technology, took advantage of Hahn's visit to Guatemala to invite him to a private meeting with Arturo Herbruger, Vice-president of Guatemala, who was also National Council for Science and Technology President and Morales' direct superior. As vice-president, Herbruger also occupied a decision-making position at GUATEL. At this meeting, Morales and Hahn explained the project's importance in detail, shared the experiences of other countries of the region, emphasized the importance of having direct connection from Guatemala to the Internet, and underlined the necessity of obtaining GUATEL's aid. The vice-president assured Morales and Hahn that he would do what he could to gain the desired support of the state telephone company. Some weeks later, Morales informed Hahn that an agreement had been reached with GUATEL.

In more general terms, these challenges reflect how much the state communications companies' position regarding the Internet's value had changed. Initially seen as a threat, the Internet now represented an opportunity. The working report presented to the Organization of American States in 1995 by Costa Rican CRNet engineers indicated it thusly:

We've had to act repeatedly as mediators among different actors: Academic Network Administrators, Telephone Companies, and the institutions participating in academic projects. [S]een by the Telephone Companies since very recently as a curiosity or marginal activity, [the Internet] has suddenly been transformed into an incredibly important and strategic activity. (de Téramond & Brenes, 1995. p. 4)

The agreement signed between the National Council for Science and Technology and GUATEL consisted of a discount on the use of the satellite link to connect to the Internet and two years of subsidized connection for the eight academic organizations affiliated with MayaNet, in other words, the five universities, two research institutes, and the council. At the beginning, according to Furlán, GUATEL's authorization of the connection excluded email use.

In April 1995, Hahn and de Téramond traveled once again to Guatemala to settle on connection details. Their report of the activity again shows the difficulties of finding technical solutions with GUATEL. Some months later, in August, during another visit by de Téramond to Guatemala organized by Magaly Morales, the panorama was more encouraging. The report notes: "the proposal [by de Téramond and the OAS RedHUCyT project] was presented to [...] GUATEL and approved, to the great satisfaction of the national council employees and the participating MayaNet member institutions" (de Téramond & Brenes, 1995, p. 9). The next day, representatives from institutions linked to MayaNet met with de Téramond to coordinate purchasing the necessary equipment.[5]

Dedicated Internet access in Guatemala was finally achieved on December 6, 1995. Once the link had been established, aid from the Organization of American States made the consolidation of MayaNet possible. Magaly Morales, National Council for Science and Technology, found a way to fund this development. Given that Guatemala owed several payments to the OAS, Morales suggested that Guatemala could resolve its debt in exchange for being able to use all available OAS funds to purchase technology. Thus, these OAS funds permitted the acquisition of a satellite antenna, necessary equipment, and training. Project collaborators participated in several courses taught in Latin America and training workshops organized by Internet Society to establish the connection.

To propel network development, the council appointed Furlán director of the academic network and created a Board of Directors made up of one representative from each of the five member institutions. However, the difficulties working with GUATEL did not stop with the connection. Pasch and Valdés (1997) point out that, after establishing the Internet link, GUATEL blocked dial-up remote access to the network since it considered it an example of "commercial usage" prohibited within the academic network. In spite of these difficulties, the institutions linked to MayaNet initiated an awareness-raising process that helped awaken interest in the network.

Of Networks and Lobbying: SVNet in El Salvador

The connection of El Salvador to the Internet exemplifies the processes experienced in the rest of the region's countries. A turning point for the process arrived with the participation of Rafael ("Lito") Ibarra, then-Director of Computer Science at the José Simeón Cañas Central American University (UCA), to the training course on TCP/IP protocols taught at the University of Costa Rica in April 1993 (barely a year after the end of the civil war in El Salvador). The experience in Costa Rica marked Ibarra, who, in addition to fundamentals and knowledge in technological subjects, found in the network workshop a "confluence of ideals and interests" regarding computer networking (interview with the author, August 08, 2017). Ibarra returned to his country convinced of the need to establish Internet connection, but faced with a general lack of awareness: "most people didn't even know or hadn't even heard about these technologies, or about the network of networks, which made it [even] more difficult to convince someone to support our country's efforts to get connected" (Ibarra, 2017, Par. 12).

De Téramond and Hahn visited El Salvador in August 1994, invited by the National Development Foundation (FUSADES), an NGO, to meet with those most interested in establishing the connection, including authorities from the National Telecommunications Administration (ANTEL), the country's state telecommunications company. Parallel to this, and as a way to move connection forward, Ibarra asked John Postel for TLD administration for El Salvador (.sv) and a number of IP addresses, which were granted in September 1994.

Like the rest of the region, in El Salvador the necessity for a shared connection among various universities was emphasized to lower costs and legitimize the project. Ibarra put this idea into practice by creating SVNet, a network made up of three universities, the National Development Foundation, the National Council for Science and Technology, and the state telecommunications company. The SVNet idea also allowed Ibarra to create a legal figure that could administrate the country's top-level domain (.sv). His reasons for taking this activity beyond the university were very similar to those expressed by de Téramond in Costa Rica: he was trying to give greater "neutrality" to the service's administration (interview with the author, August 08, 2017).

Another step toward establishing connection was the installation of a UUCP node in the country. With funds providing by the Ford Foundation, in December 1994, Luis Furlán, Theodore Hope and Enzo Puliatti traveled to El Salvador with that goal in mind. As opposed to the region's other countries, the node was not installed at a university but at the National Council for Science and Technology, created only a month before. The following year, in March 1995, the UUCP node in El Salvador began to connect to the Internet every midnight—the cheapest time to make an international call—and thus exchanged emails with the Internet through UUNET. Ibarra recalls that some of the "first messages were written in Russian, since some people thought that SV stood for the former Soviet Union" (2000, par. 6).

In spite of the possibilities UUCP offered, the project's ultimate goal was dedicated Internet connection. That required numerous conversations with the state telecommunications company and what Ibarra described as "more than two years of work, paperwork, lobbying, technical training, public presentations, private meetings, development and presentation of projects, and much more" (2015, par. 11). The project found an important ally in the United States Embassy, who supported the implementation of these training activities.

A final step to connect to the network was the presentation of the project to the Organization of American States. In addition to granting a budget, the OAS RedHUCyT project supported negotiations with El Salvador's state telecommunications company (ANTEL) and other member universities. The proposal to the OAS included a first Internet access phase in various organizations, including ANTEL, the National Council for Science and Technology, the University of El Salvador (UES), Don Bosco University, and the José Simeón Cañas Central American University (UCA). OAS agreed to finance the purchase of routers, servers and additional telecommunications equipment, in addition to providing technical assistance. As part of the negotiations, it was agreed that the equipment would be installed at ANTEL facilities, and the company promised to install lines designated for participating institutions, as part of the agreement with OAS. This process unfolded parallel to discussions of the uncertain future of ANTEL regarding imminent privatization of the country's telecommunications sector (described in more detail in the next chapter).

In August 1995, de Téramond traveled to El Salvador to fine tune technical and administrative details with the state telecommunications administration and SVNet members. During the meeting, "we arrive[d] at a consensus concerning the equipment that [would be] acquired under the RedHUCyT project and sen[t] as soon as possible to the OAS Washington office" (de Téramond & Brenes, 1995, p. 13). Three months later, Ibarra and a group of engineers finally connected to Internet at an ANTEL site located in downtown San Salvador (known as "*Central Centro*"). To test the connection, data packets were exchanged with servers from the project in Costa Rica. Several years later, Ibarra described it thusly:

> A group of Salvadorian professionals and technicians, mostly employees of the state telephone enterprise, found themselves in a modest facility, decorated with cables and strange devices, performing tests and adjustments to the link that would constitute, back then, the country's first Internet point of presence. (Ibarra, 2000, par. 1)

The network's official inauguration was held seven months later, at the end of July 1996, at a hotel in the Salvadorian capital. As an inauguration act, Ibarra repeated an activity carried at the April 1993 course at the University of Costa Rica, which had inspired his connection crusade: "[At] that UCR event, Guy [de Téramond], with a keystroke, sent a packet to Paris and [it came] back. And here [in El Salvador] the same thing happened: from here to Paris. That was the grand inauguration" (interview with the author, August 08, 2017). Hereby, the same symbolic act both launched the regional interconnection project and gave it a certain sense of closure, as the last country in the region to connect directly to the Internet. Figure 5.2 presents a timeline of the main events and initiatives that resulted in the connection of Central America to early computer networks.

The **OAS RedHUCyT** project also coordinated with the National Council for Science and Technology and the National Telecommunications Administration on the implementation of a second phase of the project, which included the network's expansion to numerous institutions, including libraries. The equipment acquired with these funds arrived at the start of 1996, which permitted the consolidation and growth of SVNet. In addition to offering email, both workshops and training sessions were also offered on the use of some of the first applications available on the network (Gopher, Veronica and Archie).

As stated above, the Web experienced an early boom in the region starting in 1995. In the case of El Salvador, the first website in the country came into existence in March 1996: the site of the José Simeón Cañas Central American University. Ibarra created a column in a local newspaper

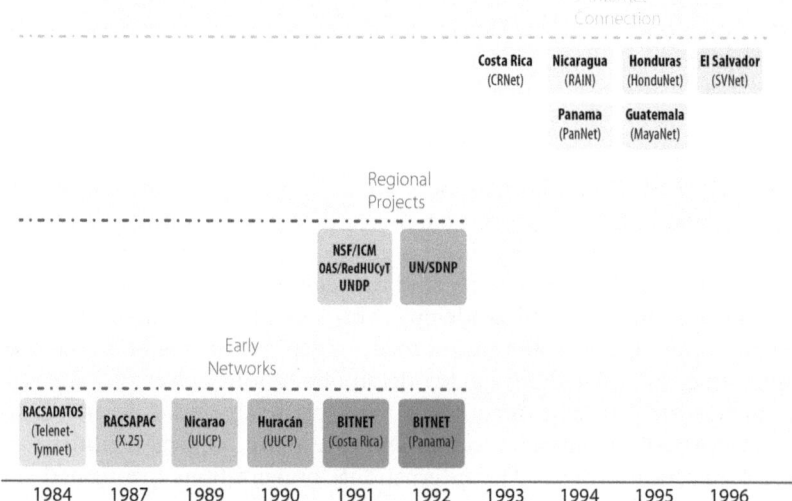

Fig. 5.2   Timeline of key networking initiatives in Central America

Table 5.1   Number of nodes, servers, and domains in Central America (1996)

Country	Nodes	Web servers	Domains
Costa Rica	5354	346	422
El Salvador	234	53	79
Guatemala	1087	112	184
Honduras	872	157	134
Nicaragua	913	107	126
Panama	1338	32	55

Source: CRNet

to discuss Salvadorian websites and thereby "inform about Internet culture" (interview with the author, August 08, 2017).

Table 5.1 reflects the state of the Internet connection in Central America in 1996, the year in which the six countries were connected, according to estimates by CRNet.

In addition to these numbers, new verbs that appeared, such as to "surf" and "navigate" the Web, captured the emergence of new ways of using technology that were normalized in specific sectors of the Central American population. The expectations of promoters were thus echoed by regional network users.

Concluding Remarks

From 1993 to 1996, Central America finally connected to the Internet on a permanent basis. This process combined the work and ambitions of local actors and initiatives, international organizations, and "network entrepreneurs." The local projects, in most of the region's countries, began due to the motivation of several academics who sought to maintain means of communication with colleagues abroad. In general, the region's countries opted to create collaborative networks among various universities and academic institutions (and some NGOs) in order to share the elevated costs of an eventual satellite connection and give more political legitimacy to the connection project. The international organizations maintained the type of work discussed in the previous chapter, that is to say, organizational efforts among interested actors, fund distributions to resolve the material needs to connect to the network, and political negotiation with governmental and institutional actors in each Central American country. Finally, "network entrepreneurs" (most notably Hahn and de Téramond) acted as intermediaries between other actors and were vital in the circulation of individuals, knowledge, and technologies during those years. In that sense, they were the perfect allies of local actors and international organizations. They carried on their shoulders (and in their suitcases) the organizational and negotiation efforts necessary to implement the network in the region in the midst of a complex technological, institutional, and political panorama that characterized the end of the 80s crisis and the onset of telecommunications privatization.

The focus adopted in this chapter also helped to better understand the consequences of transnational processes for the way in which the Internet was implemented at a local level. In each country, transnational exchanges shaped local projects in particular ways. Transnational ties strengthened de Téramond's work in Costa Rica and allowed the country to become a catalyst for the Internet in the region. The phrase used to describe the connection between Costa Rica and Nicaragua by way of a microwave link—in other words, the "first pure Internet link" in history (Arce & Hopmann, 2002)—could also be applied to refer to the exchanges among these countries' actors that led to this material connection. In Panama, the Internet represented an extension of the links that first gave rise to BITNET. The projects and ambitions to make the Internet the center of a Honduran duty-free zone acquired new dimensions through transnational connections. The action capacity of international organizations and "network entrepreneurs" was indispensable in overcoming concerns of a state Internet monopoly in Guatemala. In El Salvador, the parallel projects to connect to UUCP and the Internet came about solely through the transnational relationships established by the coordinators of these initiatives. With its differences and similarities, points of agreement and disagreement, the history of the Central American Internet up to the mid-90s was transnational through and through.

As demonstrated in this chapter, the work to expand the Internet in Central America was motivated, from its beginnings, by a regionalist impetus to construct a Central American network (in the words of de Téramond, a regional networking "backbone"). The following chapter gives a more detailed account of the motivations and obstacles of this regionalist project.

Notes

1. In addition to the countries analyzed in this work, de Téramond's task force consulted on or participated in the connection of countries such as the Bahamas, Belize, Jamaica, and Trinidad and Tobago, among others.
2. Chronologically, this process to privatize telecommunications was brought about on the following dates: Guatemala (1998), Panamá (1997), El Salvador (1998), Nicaragua (2001), Honduras (2003), y Costa Rica (2007).
3. According to Hopmann, this required the aid of UUNET and of several German academic institutions, given the American intervention in Nicaragua. These entities acted as intermediaries so that the United States Department of Defense (DDNMIL) would give the TLD to UNI (Arce & Hopmann, 2002).
4. This is an ongoing controversy that has persisted up to the present.
5. The agreements made in this meeting were the subject of dispute in the subsequent months among several members of MayaNet and the OAS project.

References

Abbate, J. (2000). *Inventing the internet.* Cambridge, MA: MIT Press.
Arce, M. E., & Hopmann, C. (2002). *eReadiness and requirements fundamentals and objectives for the CDG of Nicaragua.* Unpublished work.

Bogarín, R. (1994a). *Descubra el mundo de Internet*. Cartago, Costa Rica: Editorial Tecnológica de Costa Rica.

Bogarín, R. (1994b). La telemática en nuestro país de cara a la era de la información. In *Ideario de la ciencia y la tecnología: Hacia el nuevo milenio* (pp. 87–97). San José, Costa Rica: Ministerio de Ciencia y Tecnología.

Bull, B. (2005). *Aid, power and privatization: The politics of telecommunication reform in Central America*. Cheltenham, UK: Edward Elgar.

Burt, R. S. (2000). The network entrepreneur. In R. Swedberg (Ed.), *Entrepreneurship: The social science view* (pp. 281–307). Oxford, UK: Oxford University Press.

Cerdas, M. (1994). Información para una capital de la información. In *Ideario de la ciencia y la tecnología: Hacia el nuevo milenio* (pp. 15–20). San José, Costa Rica: Ministerio de Ciencia y Tecnología.

de Téramond, G. (1993). Description of the research network initiative in Costa Rica. *Internet Society News, 2*(1), from http://ftp.funet.fi/index/ISOC/isoc_news/issue2-1/issue2-1-complete.txt

de Téramond, G. (1994a). Interconexión de Costa Rica a las grandes redes de investigación Bitnet e Internet. In *Ideario de la ciencia y la tecnología: hacia el nuevo milenio* (pp. 61–86). San José, Costa Rica: Ministerio de Ciencia y Tecnología.

de Téramond, G. (1994b). A year in the life of CRNet. *Internet Society News, 2*(4), from http://www.nic.funet.fi/index/ISOC/pub/isoc_news/2-4/complete_issue.txt

de Téramond, G., & Brenes, A. (1994). *Establecimiento y consolidación de los proyectos Internet en Centro América y el Caribe bajo el marco del proyecto RedHUCyT*. San José, Costa Rica: CRNet.

de Téramond, G., & Brenes, A. (1995). *Establecimiento y consolidación de los proyectos Internet en Centro América y el Caribe bajo el marco del proyecto RedHUCyT. Etapa II: Hacia el establecimiento de un backbone regional*. San José, Costa Rica: CRNet.

de Téramond, G., Brenes, A., Espinoza, D., & Bonilla, R. (1997). *Establecimiento y consolidación de los proyectos Internet en Centro América y el Caribe bajo el marco del proyecto RedHUCyT. Etapa IV: Extensión de los proyectos e interconectividad regional*. San José, Costa Rica: CRNet.

Fonseca, C. (1994). La telemática: Nueva dimensión para el desarrollo educativo. In *Ideario de la ciencia y la tecnología: Hacia el nuevo milenio* (pp. 121–140). San José, Costa Rica: Ministerio de Ciencia y Tecnología.

Furlán, L. (2007, January 9). *Guatemala: Una pequeña historia de Internet*. Retrieved from https://interred.wordpress.com/2007/01/09/una-pequena-historia-de-internet-en-guatemala/

Gillespie, T. (2006). Engineering a principle: 'End-to-end' in the design of the internet. *Social Studies of Science, 36*(3), 427–457.

Hopmann, C. (1998, March 23). *Internet en Nicaragua: La historia de las oportunidades desaprovechadas.* Retrieved from https://interred.wordpress.com/1988/03/23/internet-en-nicaragua-la-historia-de-las-oportunidades-desaprovechadas/

Hughes, T. P. (1994). Technological momentum. In M. R. Smith & L. Marx (Eds.), *Does technology drive history? The dilemma of technological determinism* (pp. 101–113). Cambridge, MA: MIT Press.

Ibarra, R. (2000). Historia de la Internet. *SVNet.* Retrieved from http://www.svnet.org.sv/svnet.php?url=historia.php

Ibarra, R. (2015, mayo 10). Una noticia salvadoreña de 1995 que no llegó a los medios de comunicación. *Blog de Tecnología.* Retrieved from http://blogs.laprensagrafica.com/litoibarra/?p=3428

Ibarra, R. (2017, 22 enero). Historia de Internet en El Salvador (2a entrega). *Blog de Tecnología.* Retrieved from http://blogs.laprensagrafica.com/litoibarra/?p=4149

Law, J. (1987). Technology and heterogeneous engineering: The case of portuguese expansion. In W. E. Bijker, T. P. Hughes, & T. Pinch (Eds.), *The social construction of technological systems: New directions in the sociology and history of technology* (pp. 111–134). Cambridge, MA: MIT Press.

MICIT (Ed.). (1994). *Ideario de la ciencia y la tecnología: Hacia el nuevo milenio.* San José, Costa Rica: MICIT.

Pasch, G., & Valdés, C. (1997, February 3–5). *The dawn of the internet era in Guatemala.* Paper presented at the international federation of information processing, Florianopolis, Brazil.

Siles, I. (2008). *Por un sueño en.red.ado. Una historia de Internet en Costa Rica (1990–2005).* San José, Costa Rica: Editorial de la Universidad de Costa Rica.

Siles, I. (2012). Establishing the internet in Costa Rica: Co-optation and the closure of technological controversies. *The Information Society, 28*(1), 13–23.

Siles, I. (2017). 25 years of the internet in Central America: An interview with Guy de Téramond. *Internet Histories, 1*(4), 349–358.

CHAPTER 6

Internet and Integration in the Era of Privatization

Abstract This chapter discusses how the privatization of telecommunications took place in each country of Central America and its implications for thinking about development issues. In this context, new Internet access providers emerged. This shift profoundly modified the conditions under which the first networking initiatives operated and the ways in which computer networks evolved in Central America. The chapter shows how achieving regional integration through computer networks lost meaning for actors who turned to local solutions to resolve new technological challenges. The chapter is devoted to dissecting the different aspects of the privatization of the telecommunications field and the tensions it generated for the posterior development of the Internet in Central America.

Keywords Integration • Internet service providers • Privatization • Telecommunications

During the second half of the 90s, computer networking development followed different courses in each country of the region. Once local connection in each country had been established, regionalist impetus began to fade. Various initiatives emerged to continue regional connection efforts, but with limited results. To understand why integration through Internet technologies was not a greater success, it is necessary to analyze

© The Author(s) 2020
I. Siles, *A Transnational History of the Internet in Central America, 1985–2000*, Palgrave Macmillan Transnational History Series, https://doi.org/10.1007/978-3-030-48947-2_6

how the institutional Internet field was reconfigured during the second half of the decade in Central America.

One part of this institutional transformation related to reconfigurations of the telecommunications sector, which was shaped by major privatization efforts. As the market of telecommunications was opened for private intervention in all of the region's countries (except Costa Rica), the Internet left the academic terrain in which it had emerged. Different types of Internet service providers appeared to offer a commercial access to the network. This new institutional context put a considerable strain on the region. On the one hand, initiatives to find more network users emerged, as did efforts to promote public policies to encourage Internet access in the most disadvantaged sectors (these efforts typically focused on promoting connection through "telecentres"). Also, the first precedents were established for a Latin American theory to understand the relationship between the Internet and the region's development.

On the other hand, opening up the telecommunications market put the pioneering connection projects of universities and nongovernmental organizations at a crossroads that implied a redefinition of their goals. The idea of connecting the region *through* computer networks began to lose pertinence for actors who could turn to new access providers to connect *to* the Internet at better rates. This chapter focuses on dissecting the different aspects of this institutional transformation and the tensions that it generated.

Opening the Telecommunications Market in Central America

A point of departure for the course that Central American network connections took during the second half of the 90s is the privatization or opening of the telecommunications market. Different models were implemented in the region in the 90s, as part of a wave of opening markets in Latin America and other parts of the world. This variety of adopted models can be explained by the convergence of several factors: the pressure of international financial entities to reduce the size and functions of the state; the role of local elites in the private sector; and macroeconomic concerns, such as the desire to improve fiscal conditions (Bull, 2005; Hoffmann, 2004; Raventós, 1998). The early rise of the mobile telephone in the region, offered more broadly in countries of the isthmus since the start of the 90s, and a general dissatisfaction with the services offered by public telecommunications enterprises (except in Costa Rica), also impacted the turn towards privatization (Belt, 1999; Pasch & Valdés, 1997; Raventós, 2001; Rivera, 2007).

The arrival of the Internet coincided with a significant transformation in the value of telecommunications companies for Central American governments. Bull argues that this explains the success of local elites in their desire to participate in a field that had traditionally been directed by governments and, in the case of Guatemala and Honduras, with heavy military interference. As Bull (2005) notes, after the change from considering telecommunications to be a strategic high-impact sector for national security to a sector that promised significant profit potential for the private sector, the elites from the latter sector became the most central actors (p. 26). Against that backdrop, the telecommunications market's privatization was promoted as a solution to management issues, and as an opportunity to reach the potential generated by this field for the development of the isthmus.

In general, the processes of negotiating and implementing privatization policies were not particularly transparent in most of the region's countries and were characterized by numerous irregularities and, for some of them, accusations of corruption, notably in Guatemala (Pasch & Valdés, 1997). Although different models were used in practice which resulted in relatively disparate organizations, opening up telecommunications followed three major logics or "points of departure" (Bull, 2005, p. 35).

Guatemala and El Salvador implemented a competitive model with very similar legislation and received ample support from the United States Agency for International Development (USAID) for their privatization process. This agency not only offered economic aid in exchange for approving privatization reforms, but it also provided training for its implementation. For example, it financed trips for local operator officials and legislators to places like Chile and Berkley, to discuss in depth the details of said projects with advisors (Belt, 1999). The main focus of this privatization model was the promotion of competition in the sector. To do this, a proposal was put forth to break away from state enterprises and privatize aspects like conflict resolution and spectrum management (Argumedo, 2007; Urízar, 2007).

In El Salvador, the state enterprise was divided in two: one dedicated to landlines and the other to mobile communications. In 1998, both enterprises were privatized. France Telecom acquired 98% of landline operators and Telefónica de España (Spain's multinational telecommunications company) acquired the mobile operator (Argumedo, 2007). In Guatemala, a new corporation was created (Telecommunications in Guatemala, TELGUA) to which all assets and liabilities of the Guatemalan Telecommunication Enterprise were transferred. In 1998, 95% of TELGUA's shares were sold to the company Luca, a group of local investors. At the start of 2000, when "it became clear that Luca did not have the funds [to pay for the TELGUA purchase], Telmex [Mexican telecommunications enterprise] purchased 79 per cent of the stocks of Luca" (Bull, 2005, p. 69). In both countries, creating an independent regulatory agency with limited power was proposed (although in practice this did not succeed) (Urízar, 2007). Several mechanisms were also created to provide subsidies and, in this way, offer services in low-income areas.

Honduras, Nicaragua and Panama were the first countries in the region to approve laws aimed at opening the sector up to private participation in 1995. The proposed model in these countries promoted a more limited competitive system and a stronger role for state enterprises than in the case of Guatemala and El Salvador. These could maintain their monopoly position for set period of time as an interim measure and as an accompanying mechanism prior to liberalization. In that sense, they constituted a more representative example of the model employed in the rest of Latin America (Raventós, 1998).

Nevertheless, the trajectory of these processes followed somewhat different paths. In Panama, the reform was implemented in 1997. A period of exclusivity was given to the National Telecommunications Institute, but only for landline management. For mobile communications, participation was limited to two operators, and 51% of the state telecommunications company's shares were sold (González, 2007). Given the state of their economies and their high foreign debt, Honduras and Nicaragua were particularly influenced by international financing entities during their privatization processes. However, and in spite of this pressure, the process was slower and more complicated than expected. By 2000, and after three failed attempts in both countries, privatization had not been achieved in either of the two (Ansorena, 2008; Tábora, 2007). Raventós (2001) argues that to resolve the problems generated by the "piñata" (the acquisition of a group of properties by Sandinistas before the change in government) was an important motivation for the Violeta Chamorro administration in its desire to privatize. During the process of Nicaraguan privatization, assets were transferred to a new state company (the

Nicaraguan Telecommunications Enterprise, ENITEL), and the Nicaraguan Institute of Telecommunications and Postal Services stayed on as a regulatory entity. Privatization was finally achieved in 2001, when 40% of the state company's shares were sold to a Swedish-Honduran consortium (at a much lower price than was expected by the government) (Rivera, 2007). In 2004, the state sold the remaining shares. In general terms, these countries promoted the operation of state regulatory entities, which would be responsible for regulation and spectrum management, as well as rate fixing and conflict resolution.

Over the course of the 90s, Costa Rica was the only country in the region that did not open the telecommunications market to private participation. This did not occur until 2007 and required the first referendum in the country's history, which voted in favor of privatization by a slim margin. This does not mean that privatization did not have its promoters in the country. Proposals to partially open up the sector had been negotiated since the Oscar Arias administration in 1986. In August 1996, the Legislative Assembly received an array of reforms and proposed bills to that end, known as the "ICE Combo." The proposed bills were discussed over the following years, until their approval in the first Legislative Assembly debate in the early 2000s.

The "Combo" proposed dividing the Costa Rican Electricity Institute (ICE) into two specialized enterprises (one in electricity as a public utility and the other in telecommunications), which would remain in the hands of the state, and a gradual opening of the market: first the value-added services (including the offer of Internet access), then those of telecommunications and, finally, telephone services.

The proposal was defended by its promoters as an attempt to modernize, improve service provisions, and reduce public investment in the telecommunications sector (Monge, 2000). It was about, then, a "very conservative [model] involving a very slow liberalization without privatization" (Raventós, 1998, p. 1). The continuation of "universal access" to technology was guaranteed through the creation of a special fund.

This plan for opening access was met with fierce opposition from civil society, public universities, social movements, and unions. This opposition can be understood as the defense of an institution that constituted an emblem of Costa Rica's development model and democracy (Haglund, 2006; Hoffmann, 2004, 2008). The infrastructure developed by the Costa Rican Electricity Institute over several decades was also one of the best in Latin America and had been decisive in the development of a successful process of attracting direct foreign investment that contributed to the installation of an Intel assembly line factory in the country in 1998 (Fumero, 2013; Raventós, 1998; Siles, Espinoza, & Méndez, 2016). A result of this failed privatization process was the formation of a "joint commission" to discuss the reformulation of the bill, which led to a series of recommendations to strengthen the Costa Rican state telecommunications operator (ICE).

A key factor in the operation of the Central American telecommunications field (and market) was the action capacity of the new regulatory agencies. In practice, the autonomy of these agencies was relative. On the one hand, they had to assume their duties when the sector opened up to private participation, which put them at a disadvantage (Rivera, 2007). On the other hand, they had to face the challenge presented by the same local elites who had promoted privatization in the region (Bull, 2005). Finally, the laws that authorized the access opening established few mechanisms for regulating competition (Argumedo, 2007; Tábora, 2007).

Toward a Commercial Internet

In addition to the privatization of the telecommunications sector, a second element necessary to understand the reconfiguration of the institutional Internet field in Central America relates to the network's exit from predominantly academic spaces and the emergence of commercial-access providers. In most of the countries, this occurred in the context of the sector's opening (1994–1995). The first commercial provider in the region was Costa Rican Radiographic Institution (RACSA), subsidiary of the Costa Rican Electricity Institute that had implemented the X.25 network (RACSAPAC), still in operation in the monopoly framework (see Chap. 3). To comprehend how these types of providers emerged and the implications of their operation, in what follows the case of the Costa Rican Radiographic Institution and its gradual transition to the Internet is examined.

From X.25 to the Internet: A Forced Shift

At the start of the 90s, the Costa Rican Radiographic Institution (RACSA) had focused on operating and strengthening the use of X.25 at a regional level. Considering future options, the company was torn between recognizing that X.25 had reached its end, and therefore it would be worth considering new options like the Internet, and wanting to keep promoting its use given the investment made and the degree of specialization acquired during its operation (Aguilar & Ballestero, 1993). In that sense, given the historical emphasis of RACSA on the country's business sector, the goal was to explore the computer networking field at the corporate level through Frame Relay. This technology provided several solutions to the use of X.25 and, hence, it was announced as the next step in the history of network development at the company.

However, the growth of the Internet presented a series of challenges for institution. At an international level, the Internet had gradually won the battle of becoming the standard (Abbate, 2000). Therefore, the U.S. central offices and those of corporations operating in Costa Rica had already begun to explore the option of the Internet. At a local level, the University of Costa Rica project had attracted some X.25 clients (such as BBS and public database creators, for example). The sum of both factors threatened to leave the Costa Rican Radiographic Institution with an expensive yet underutilized network.

These factors were compounded by a growing interest in the Internet from engineers and collaborators within the company. Gabriela Guido (the executive who worked with the Costa Rican National Research Network, CRNet, at RACSA), Juan Carlos Blanco and Álvaro Retana (from the Telematics Development Department) and, later, Gonzalo Berrocal and Mario Serrano (from the computer science area), stood out among the interested in the project. The root of this interest in the Internet varied. Several new collaborators from the company had entered after the X.25 boom, so their attachment to this technology was not as strong as that of other colleagues who had been at the Costa Rican Radiographic Institution longer. Others had been affiliated with CRNet, either as students at the country's public universities or through their work itself at the institution. Finally, Álvaro Retana had studied electrical engineering in North Carolina before joining the company. In that context, he had access to academic networks like the Internet.

Insomuch as the interest in the Internet began to consolidate, this group of collaborators tried to promote connection to the network at the company. This challenge had two components. On the one hand, it was necessary to overcome the skepticism with which some engineers viewed the Internet. Regarding this, Gabriela Guido comments:

Breaking the paradigm was not easy because there was a natural resistance to change. Although it was known that the X.25 had already reached its end it still wasn't clear what the next step to follow was. [...] One of the biggest questions about the Internet and its protocols for the engineers that saw the X.25 come into the world and grow was who would do the invoicing and who would handle the correspondence. They asked me constantly who would manage the Internet, pay for its maintenance, and who would RACSA [the Costa Rican Radiographic Institution] have to pay to use it. The process of convincing these engineers and the rest of the staff wasn't easy, since many years had been spent operating this one technology. (Interview with the author, September 19, 2006)

Additionally, the company's highest-ranking officials had to be convinced. As a result of conversations with the company CEO (particularly the assistant manager Carlos Moreno), in 1993 the Costa Rican Radiographic Institution authorized a pilot project for commercial Internet access. In the words of Juan Carlos Blanco, one of the engineers involved in its development, this was "more of a proof of concept than an implementation project. At that point, [the] Internet started to take off, and the business started to grow at [the institution]" (interview with the author, March 07, 2006). To drive the project forward, the team at Costa Rican Radiographic Institution resorted to academic connection projects. From 1993 to 1994, CRNet engineers provided technical assistance to the staff in charge of the pilot project at RACSA.

Presentations to the company's management and different internal departments accompanied the preparation process. Álvaro Retana remembers that graphics were used that demonstrated the international growth of the Internet worldwide (interview with the author, March 15, 2006). In regard to a presentation given to some 40 collaborators from different RACSA departments, Guido recalls:

We went [into] MIT [Massachusetts Institute of Technology] and to the U.S. Library of Congress. At that time, we would see a Unix "prompt", a ">", we would type "telnet" and the name server, and a window (also in text) immediately popped up indicating that we were already at the host's access gateway. (Interview with the author, September 19, 2006)

An influential factor in the management's decision to encourage the Internet project was the availability of graphic interface applications. These were the first versions of Internet browsers that, although in an emergent state, showed possibilities that were not commercially available on other computer networks. Even with certain reservations generated by the economic cost that financing the project on a larger scale would imply, and the uncertainty surrounding the return on investment that could be expected, the pilot plan completed its mission and ended up convincing high officials at the Costa Rican Radiographic Institution. The company began to acquire its own equipment to move forward with opening the service. The commercial Internet connection project was organized based on three working areas: network routing; the creation of a market for the network; and the computer science area, dedicated to the configuration of applications like email, among others.

The year 1994 marked the beginning of the commercial offer of Internet access in Central America. Over the course of that year, the service began to gradually take off. In Costa Rica, among the first clients of the Costa Rican Radiographic Institution, enterprises related to services stood out, along with tourism, computer, diplomatic and academic organizations, and even nongovernmental organizations. The growth in demand implied a gradual process of designing and expanding the network, requisition and procurement of equipment, developing the market, and training for the team assigned to the project. The company offered the options of dial-up access and dedicated connection. Employees of the company visited corporations that were interested in the service to demonstrate how the network worked. In addition, enterprises with close ties to the company were named "authorized agents" in promoting this Internet-access service.

By 1995, the number of clients surpassed, for the first time, the number of internal users at the company, which represented for promoters "a landmark in the history of the Internet at the Costa Rican Radiographic Institution" (Mario Serrano, interview with the author, August 10, 2006). In addition to email, the rise of the Web helped consolidate the institution's Internet project. Thus, the notion of a "Graphic Internet" started to be promoted within the company, a commercial name given to the Web. From 1994 to 1996, the commercial Internet-access network maintained growth. By 1999, the enterprise had almost 29,000 clients in the country's residential access sector, compared to the 300 in 1996 (Siles, 2008). In that context, the X.25 network was definitively closed in 1999. By that time, Internet access services represented almost half of the institution's total income, a tendency that would be sustained over the coming years. At the start of the new century, the company began to discuss the possibilities of amassing a hundred thousand clients, a plan that was known as the "Internet of 100,000" (Alberto Bermúdez, interview with the author, March 15, 2018).

New Providers, New Tensions

The Costa Rican Radiographic Institution was the first of several commercial Internet access providers that emerged in Central America. Private operators began to offer services in Guatemala, Nicaragua and Panama between 1994 and 1995. In 1996, it was estimated that more than 40 organizations (public and private) offered network-related services (Pasch, 1996). By 1997, the number of service providers had risen significantly in countries such as Honduras and Guatemala (Pasch & Valdés, 1997; SDNP, 1997). These providers promoted a series of similar services: email, file transfers, Gopher, and the emerging Web. In general, the first network users were concentrated in capital cities (Sáenz & Galeano, 1996). Newspapers like *La Nación* (in Costa Rica) and *Gerencia* (from the Guatemalan Management Association) became two of the first to develop their own Web version (Pasch & Valdés, 1997; Siles, 2008). In that way the first precedents were established for the formation of an emerging regional Internet industry.

The arrival on the scene of private service providers involved a series of negotiations with state telecommunications operators within the framework of the region's gradual market opening. Pasch and Valdés (1997) describe the requirements originally established by the Guatemalan Telecommunications Enterprise for Guatemala's new operators: "First, every ISP was forced to transfer to GUATEL 10% of all Internet related sales, including services, Internet seminars, and even 'revenue derived from the use of GUATEL's logo,' plus 15% of related sales of software, and 4% of modem sales" (p. 9). A common initiative to face this type of negotiation was to form groups of new access providers, which were organized in every country in the region (except Costa Rica) to gain legitimacy and negotiation power.

The emergence of commercial Internet access providers also signified a major adjustment in the operation of the academic organizations that had led the network's development since the mid-90s (analyzed in Chap. 5). Inasmuch as the telecommunications market opened up to private participation and new access providers appeared, the academic Internet connection projects lost impetus. Various universities sought to resolve their own networking necessities individually and no longer as part of a network of academic actors. Luis Furlán refers to the case of Guatemala with words that could describe the situation in other countries in the region (except Costa Rica, where the academic network maintained a significant activity until the start of the following decade): "In 1998 [there were] already several Internet Service Providers in the country and their rates were lower than those offered by MayaNet. Each university hired whichever ISP services seemed best, and that was the end of the Academic/Scientific MayaNet network" (Furlán, 2007, par. 19).

Given that commercial access providers could resolve individual connection problems at increasingly lower costs, operating a strictly academic network dedicated to facilitating the use of the network started to lose relevance for many of its members. In the following years, support in the framework of projects such as the OAS's Hemisphere-Wide Inter-University Scientific and Technological Information Network (RedHUCyT) were reduced to mainly long-distance consulting and several visits to the region's countries (de Téramond, Brenes, Espinoza, & Bonilla, 1997).

A similar situation characterized the development of the first NGOs that had promoted the Internet (and, before that, UUCP) in the region's countries. The projects of the Sustainable Development Network faced the same difficulties that put the academic networks in check. The trajectories of these Sustainable Networking Development Programme projects are revealing in that they reflect the state of the region's emerging Internet industry during the second half of the decade. In Honduras, the country of the region where the SNDP project had experienced the most success, those in charge of the project opted to transform it into an organization, described thusly:

An independent, impartial, apolitical, and nonprofit nongovernmental association that seeks to foster sustainable development using Information and Communications Technology (ICT) such as tools to manage, share, and spread information resources and strengthen institutional and communitarian capacities. (RDS, s.f., par. 4)

Guatemala's SNDP project, established in 1997, chose a similar solution:

> Develop and maintain a competitive edge in this rapidly evolving ISP market by focusing on service quality, lower prices, offering technical training, and specialized priority sectors [...]: education, environmental legislation, biodiversity, health, energy, solid waste management (SDNP, 1997, pp. 24–25)

In Panama, academic efforts were redirected toward an interconnection project different from network service provider operations. At the initiative of the National Secretary of Science, Technology, and Innovation (SENACYT), with financial aid from the OAS RedHUCyT project, and technical assistance from the Network Startup Resource Center (NSRC), in April 1997 "Intered" was inaugurated, the first Internet Exchange Point (IXP) in Latin America. The IXP made it possible for local access providers to exchange data with other providers, without said data leaving Panamanian territory, as was the norm up to that time. Originally, the project interconnected five providers: the academic network (PANNet) and four commercial providers (C-Com, GBM, OrbiNet and SinfoNet).

Randy Bush and Dave Meyer, Network Startup Resource Center collaborators who participated in the IXP's technological configuration, explained the initiative's success in terms of six factors: an emerging Internet service offer industry, the impossibility of monopolizing these services, the existence of a physical low-cost infrastructure, local knowledgeable engineers, institutional support from key organizations and "willingness to take the risk of cooperation" (Bush & Meyer, 1997, slide 4). Since the project was oriented toward articulating interests among various actors, the creation of the IXP in Panama (and, several years later, in Costa Rica) was seen as a natural continuation of the pioneering role of academic institutions in Internet connection (Hahn, 1999; Siles, 2017). To implement it, an organization governed by its own regulations and financed by collaboration from participants was created. In 1999, nine local providers interconnected through Intered (Hahn, 2000).

Costa Rica, where efforts to privatize and to open markets did not prosper during the decade, was the first country in the region where the academic network experienced the greatest prominence in the second half of the 90s. This process was shaped by significant controversies (Siles, 2012). Insomuch as the Internet demonstrated economic value, the relationship between the Costa Rican Radiographic Institution and CRNet began to deteriorate. The friction between these groups centered on the network's legal validity in the context of the national telecommunications monopoly. Authorities of the Costa Rican Radiographic Institution—the state telecommunications operator subsidiary—argued that, since these networks involved the use of telecommunications infrastructure, the company had to become the country's sole Internet service provider. From that perspective, although the academic project had been vital for installing Internet for the first time in the country, the network's commercial expansion required stricter application of monopoly laws. Those responsible for the academic network, on the other hand, maintained that the restrictions on their work would lead to significant obstacles for the country's scientific development.

In addition to legal aspects, this controversy also involved both technological and economic matters. The imminent end of the X.25 experiment forced the Costa Rican Radiographic Institution to search for alternatives to recoup investments. Therefore, the company prioritized installing dial-up Internet access in homes at a relatively low cost. In contrast, de Téramond believed that only dedicated lines would guarantee network scalability in the long term. Additionally, he lobbied to provide free Internet access to the academic network members. In a letter published by *La Nación* newspaper in May 1996, de Téramond attributed Costa's Rica's deterioration as leader of Latin American indexes of Internet development to the Costa Rican Radiographic Institution (RACSA), and described the state company's attempt to take over the academic network:

> Insomuch as Internet-related technologies establish themselves as dominant technologies and invade the commercial sector, the Internet ceases to be a curiosity for telephone companies, and a natural tension is established between the powerful telephone monopolies and academic networks. Our country is not the exception, and in mid-1994, RACSA, surprisingly, after initiating its commercial service, communicated to the National Network its intention to take over CRNet operations, a strange situation, given that CRNet pushed RACSA for more than a year to initiate its commercial service. In light of this fact, greater efforts were needed from CRNet's board members as well as decided support from the former Minister of Science and Technology, Dr. Roberto Dobles [Executive President of the state telecommunications operator at that time], and Eduardo Sibaja [Viceminister of Science and Technology] to sustain the network. (de Téramond, 1996b, Par. 8).

The frictions between commercial and academic networks were widely described in a September 1996 letter sent by PanAmSat representatives to the president of Costa Rica and other political figures, in which they responded to the Costa Rican Radiographic Institution's formal request to disconnect CRNet. In this letter, PanAmSat called upon Costa Rican authorities to overcome the "nontechnical difficulties" between both projects and find a solution that would not risk access of the "Costa Rican educational community to the Global Information Superhighway." CRNet was able to continue service due to the intervention of the Minister of Science and Technology.

The distance between the academic project and the Costa Rican Radiographic Institution grew during the second half of the 90s. Both groups established alliances with new actors to justify its models and work principles. In April 1997, for example, CRNet inaugurated a satellite station donated by the Organization of American States to the University of Costa Rica. Through this station, the academic network's link to the Internet was channeled without utilizing monopoly infrastructure.

A Network in Search of Users

Inasmuch as issues of infrastructure and access were being resolved, Central American Internet promoters worked to find more users for the network. Commercial access providers designed media campaigns primarily focused on higher-income sectors, as a way to recoup the investment generated by acquiring Internet infrastructure. In countries like El Salvador and Guatemala, creating "social development funds" was considered in the framework of privatization and opening access to guarantee access to the technology in low-income areas. However, its scope was limited (Sáenz & Galeano, 1996). As stated above, several NGOs (notably the Fundación Acceso in Costa Rica and the Sustainable Development Network in Honduras) focused on offering Internet-related services (such as website design, and creating discussion forums about relevant topics) to other local or regional organizations.

The icons of the 1990s for Central American Internet access were "cyber cafés" and "telecentres." Proenza, Bastidas-Buch, and Montero (2001) summarize the characteristics of these places: "It consists of premises stocked with several computer terminals and simple furnishings [...] The main service offered to the public is access to the Internet (chatting, e-mail and Web browsing) and often also to elementary software (word processing, spreadsheet)" (p. iv).

In spite of their similarities, telecentres and cybercafés revealed relatively different political views of Internet access. The telecentre, typically financed by governments or nonprofit organizations, made the developmental expectations offered by public, collective, and free access to the network clear (Gómez & Martínez, 2001a). In that sense, telecentres embodied what promoters envisioned as "a judicious development strategy that is in accord with the resource-poor situation of developing countries and marginal areas" (Proenza, Bastidas-Buch, & Montero, 2001, p. iv). The cybercafé, on the other hand, was a commercial profit-based project, usually equipped with better technological conditions than telecentres and located in places that were attractive for potential users. Further to resolving access difficulties, cybercafés and telecentres provided Internet use with a cultural mystique.

At the turn of the century, state programs for promoting telecentres were developed in every Central American country: "Municipal Telecentres" in Costa Rica, "Infocentres" in El Salvador, "Digital Centers" in Guatemala, "Multipurpose Centres" in Honduras, "Telecentres" in Nicaragua, and "Infoplazas" in Panama (Gómez & Martínez, 2001a). However, the cybercafé (more than the telecentre) was the one that became a common network access option for a considerable number of people in the region's countries who did not have connection in their homes (Monge & Hewitt, 2004). For many, this signified an obstacle to generating equitable conditions for technology access.

On Digital Agendas and Public Policies

Regional efforts to transform individual initiatives into public policies aimed at encouraging Internet connection were feeble (if not nonexistent) and dissimilar. These early efforts, shaped by the rise of the Internet and mobile phones, were discursively framed as promoting "information societies." This term, with its respective symbolic associations to a certain understanding of progress through technology, resonated throughout the Latin American region, both in institutional and governmental projects as well as in the first academic projects on the subject (Trejo Delarbre, 1996). The premise behind this notion was that "technology is not only the product of development (as it derives from the development process), but is also, to a large extent, its engine (since it is also a tool for development)" (CEPAL, 2003, p. 1). Therefore, the question, in CEPAL's own words, was not if this was the model to be followed as path to development, but rather "what type of information society [was] desired" (2003, p. 1). The main obstacle to achieve this objective was a concept called the "digital divide," extensively cited to refer to "the dividing line between the population group that already has the possibility of benefiting from [technologies] and the group that is still unable to do so" (CEPAL, 2003, p. 7).

In Central America, the most tangible (and earliest) regional expression of this technological and developmental vision in public policy was the "Digital Agenda," a national plan presented in 2000 by the Miguel Ángel Rodríguez administration in Costa Rica (Rodríguez, 2001). One of the first proposals of its type in the Latin American region, the Digital Agenda promoted five specific actions: (1) improving the country's telecommunications infrastructure in order to develop a national interconnection platform; (2) facilitating Internet access in the country to the highest number of people possible; (3) strengthening a "new economy" based on communications technologies; (4) consolidating government services by way of the Internet through the formation of a "Digital Government"; and (5) creating appropriate laws to develop the country's communications technologies, and to develop the "new economy" facilitated by technology. Although the projects developed in the framework of Rodríguez's Digital Agenda had a broad scope, they demonstrate the way in which the premises behind notions such as the "information society" and the "knowledge economy" were assimilated in the Central American region in the public policy that began to emerge at the start of the new century.

In spite of the academic, commercial, organizational, and governmental efforts to promote network use as a developmental tool, Internet use and access in Central America was modest and unequal. Martínez (2001a) classified the region's countries into three major groups that reveal the tensions of the privatization style adopted: those countries with (a) intermediate coverage, rapid development, and high Internet concentration in certain segments of the population (El Salvador and Guatemala); (b) high coverage, rapid development, and low Internet concentration in few segments (Costa Rica and Panama); and (c) low coverage, slow development, and high Internet concentration in certain segments (Honduras and Nicaragua). Table 6.1 shows these regional differences in connection statistics.

PRIVATIZATION AND THE LIMITS OF INTEGRATION

As was discussed in the previous chapter, the advance of regional Internet connection through academic initiatives, conducted from 1993 to 1996, awoke great enthusiasm in the region among its promoters, who sought

Table 6.1   Percentage of internet users in Central America (2000–2005)

Country	2000	2001	2002	2003	2004	2005
Costa Rica	5.80	9.56	19.89	20.33	20.79	22.07
El Salvador	1.18	1.50	1.90	2.50	3.20	4.20
Guatemala	0.71	1.74	3.39	4.55	5.10	5.70
Honduras	1.20	1.42	2.60	4.80	5.60	6.50
Nicaragua	0.98	1.45	1.71	1.88	2.32	2.57
Panama	6.55	7.27	8.52	9.99	11.14	11.48

Source: International Telecommunications Union

ways to strengthen its development. Nevertheless, the institutional transformations to the Central American telecommunications field described in this chapter changed the conditions that made conceiving projects of integration through technology possible. The second half of the 90s was thus shaped by a tension between initiatives to promote Central American interconnection through the Internet and the lack of willingness and political interest to bring them about.

The "Central American Internet Backbone"

Perhaps the most significant expression of this desire to make the regional interconnection network grow was the proposed "Consolidation of the Central American 'Backbone,'" by CRNet and the Costa Rican Ministry of Science and Technology, in alliance with the OAS Hemisphere-Wide Inter-University Scientific and Technological Information Network project, presented to the Central American Bank for Economic Integration (BCIE) in July 1995. (Fig. 6.1 graphically presents the proposed interconnection while also showing the state of each country's 1995 connection projects).

Chapter 5 demonstrated that de Téramond's work as "network entrepreneur," whereby he traveled the entire region helping connect its countries to the Internet, was motivated by a latent integrationist project. During the second half of the decade, de Téramond made this project more explicit. His proposal to the Central American Bank for Economic Integration argued for "accelerated growth of Central American Internet [that] carries the urgent need on the part of Central American countries for greater encouragement of the region's telecommunications development" (de Téramond & Brenes, 1995, p. 58). Additionally, he pointed out the

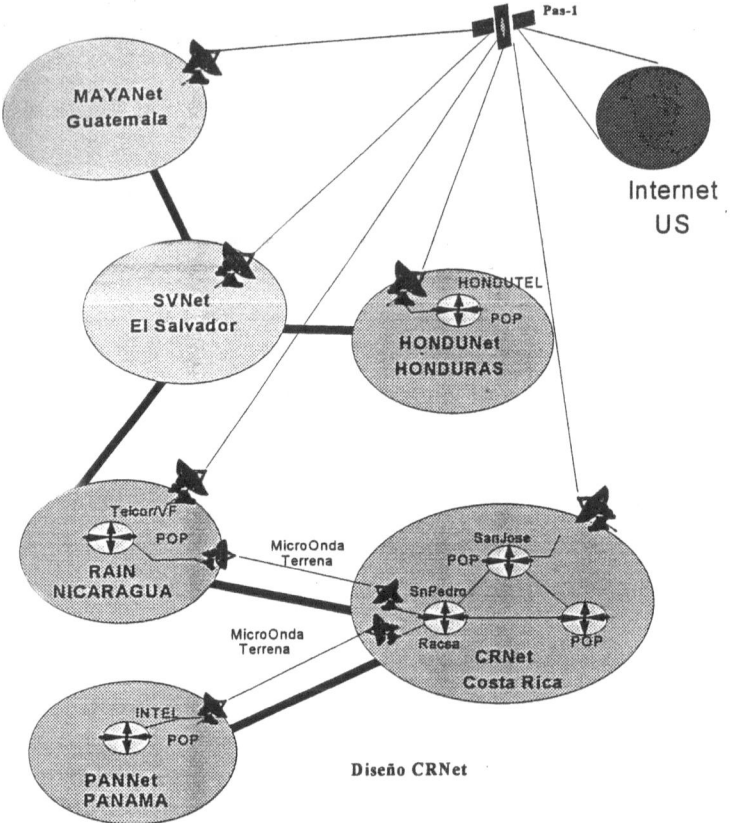

Fig. 6.1 Proposed development of a "Central American [Internet] backbone." (Source: De Téramond and Brenes 1995. Used with permission)

limitations of seeking an individual Internet link for each country, as had been achieved up to that moment, instead of regional interconnection:

> A model has been favored in which each country independently seeks its satellite Internet access. This entails the serious drawback that communications country-to-country within the region must be enabled by a double satellite signal, causing the quality of the applications in real time to deteriorate considerably, making applications like video and other interactive uses,

which are all necessary for teleconferences and distance learning, impossible in practice. (De Téramond & Brenes, 1995, p. 65)

Given this scenario, the project sought to comply with the ultimate goal of integrating Central America by way of the Internet. One of the proposal's objectives dictated: "to serve as a significant cooperative and collaborative instrument among all sectors, truly strengthening the region's integration and sustainable development processes" (de Téramond & Brenes, 1995, p. 67). In that sense, the proposal of the "Central American Internet backbone" added a developmental perspective to the vision of de Téramond over the years regarding the network as an instrument for overcoming academic isolation. Like the integrationist projects of the 60s, this development required material integration. At a conference held in 1996, de Téramond stated:

Many countries for which [the networks] are vital to their development, are kept far from the communication, information, and collaboration with colleagues of scientific institutions and major corporations, which puts their research and, in effect, their educational systems and economies, at a disadvantage. (1996a, par. 19)

In practice, this required establishing a link among all the Central American networks that had been developed from 1993 to 1996. These networks remained separate in the sense that, although they connected to the Internet by satellite, the present infrastructure was not sufficient for the countries to connect to each other. Even Nicaragua and Panama, which originally connected through Costa Rica, had developed their own satellite links. In that way, to fulfill the goal of integration through the Internet, strengthening the region's acquired infrastructure was proposed. More specifically, the goal was to both replace the analog links between Nicaragua, Costa Rica and Panama with digital higher-speed links, as well as to install digital links in the countries that had yet to be interconnected. (See Fig. 6.1). The Organization of American States, through its RedHUCyT project (presented in Chap. 4), had approved funds to integrate Panama into the isthmus' Internet backbone, given that the Central American Bank for Economic Integration funds did not cover that country.

In November of that year, the Central American Bank for Economic Integration authorized a non-refundable amount of $110,000 to finance this initiative. However, the project did not come into being, partly because reconciling the interests of the region's state telecommunications companies was difficult, and partly because the telecommunications market opening in most Central American countries changed the range of possibilities for network connection. As an example, de Téramond recalls having received a letter from the Regional Technical Telecommunications Commission (COMTELCA) authorities requesting explanation of the reasons that justified installing connections exceeding 64 Kbps among the region's countries. Integration through technology was no longer a priority for telecommunications companies, now focused on increasing their number of internal clients in the light of increasing competition.

Many Institutions, Little Integration

Other changes in the institutional Internet field made these integration processes challenging. In spite of the abundance of institutions that characterized regional integration in the 90s, there were relatively few infrastructure projects with regard to telecommunications. Meanwhile, membership to the Regional Technical Telecommunications Commission changed after the sector's privatization. Starting in 1995, representatives from regulatory agencies partially replaced local telecommunications operator delegates. Several coordination meetings were convened by the Central American Integration System (SICA) with representatives from local telecommunications operators, but the results were modest (Sáenz & Galeano, 1996).

The privatization force had significant effects on the integration processes of the 90s. Despite the service problems identified in various countries of the isthmus, state telecommunications monopolies created conditions for greater technological integration. In addition to coordinating infrastructure issues that made projects like those discussed in previous chapters possible, Raventós (1998) summarizes the benefits generated by regional collaboration: "[It allowed for] joint negotiation of settlement rates with the United states, the development of regional trunks, like the 140 Mbs microwave link, and sharing of international satellite circuits to countries with little traffic, like Switzerland, using such links" (p. 51).

Nonetheless, as the field became more and more privatized, the integrationist project practically disappeared from the agenda of local operators, who directed their efforts toward another body of local initiatives and concerns. By 1998, Raventós had already pointed out the implications of abandoning regional integration efforts: "In the future there is unlikely to be a position on settlements accepted by all countries" (p. 52). A sign of this difficulty was the impossibility to implement digital terrestrial links among the Central American region's countries.

FROM INTERNET ACCESS TO MEANINGFUL USE: ARCHAEOLOGY OF A LATIN AMERICAN ICT4D THEORY

In addition to initiatives of promotion, public policy, and emerging use cultures, the Central American Internet field included the formation of one of the first theories to think about the relationship between technology and development in Latin America. The emergence of this body of work must be situated within its proper context. To begin, it was part of an integrationist wave throughout the 90s in which the search for regional solutions to local issues was normalized (see Chap. 2). In that sense, it is impossible to separate the reflections generated in Central America from what was happening in other parts of Latin America. For instance, in mid-1999, the International Development Research Centre (IDRC) in Canada created a fund to finance research projects and activities about the topic of communications technology in Latin America and the Caribbean (Bonilla & Cliche, 2001; Gómez & Barnola, 2000). Through these types of projects and forums for exchange, a stream of reflections was created that several promoters claimed as inherently Latin American: it was the product of empirical fieldwork in Caribbean countries throughout the region; it resulted from collaborations among local and regional organizations; and it represented a series of "values," "work styles," and "strategies" that were shared and relatively typical in Latin America (Gómez & Martínez, 2001b, p. 17).

In Central America, this work was most notably accomplished at the Fundación Acceso, a Costa Rican NGO established in 1992 with the aim to strengthen the work capacities of other civil society organizations in the region. Fundación Acceso quickly placed the topic of the Internet at the center of its concerns. One of its first work programs was focused on facilitating computer network connection to Central American NGOs (including links to Nicarao and Huracán, in the days before the Internet). Insomuch as various private Internet access providers began to appear in the region, the foundation refocused its activities to strengthen the strategic use of tools like the Web.

Toward the end of the 90s, within the working framework created to encourage Internet use in Central American NGOs, a body of reflections regarding the network's value as a developmental tool started to become clear. Different conditions contributed to this process. The Acceso work team, though small, combined professionals from computer science and the social sciences. In addition to facilitating knowledge for the Web's use in Central American organizations, Fundación Acceso also conducted studies regarding the state of the region's telecommunications (Sáenz & Galeano, 1996). It was Acceso, for example, who carried out a feasibility study for the implementation of the UNDP's Sustainable Development Network in Costa Rica (described in Chap. 3). In this way, it became possible to express an understanding of the Internet's technical aspects, experiences working with organizations, and to reflect on their importance for key Central American institutions. During the 90s, Fundación Acceso also experienced relative financial stability that allowed for allocating resources and personnel to the projects. The foundation's regional presence served both to attract several donations from international organizations and to recruit collaborators from various countries. In that way, by the end of the century the foundation had a work team of more than 20 people and collaborations with entities overseas.[1]

Through its consulting work on Internet connection and Web use, as well as through its research efforts, the foundation connected Internet use to the notion of development. The basic premise of this work was that technology by itself could not be viewed as a direct connection to development (as was promoted by many international organizations). In the words of the researchers, "The Internet is not intrinsically positive or negative, though it is not neutral either. [...] [It can] reproduce and increase the social and economic inequities that exist in our societies [...] [but] we can turn the Internet into a tool for development" (Fervoy, Martínez, & Sáenz, 2000, p. 93). To do this, they argued, transcending a focus on connectivity was necessary. In its place, the researchers at Acceso promoted what they called a "social vision" of the Internet, that is to say, an approach centered on the conditions that would truly make possible maximum use of the Internet, to modify central aspects of its users' daily lives, especially those in more disadvantaged positions (Gómez, Martínez, & Reilly, 2001).

This perspective expressed a profound dissatisfaction in the lack of public policy related to telecommunications matters (notably the Internet) and in the typical orientation toward implementing "information societies" interested primarily in providing connectivity, but inattentive to the conditions that would make its access possible to encourage regional development (Martínez, 2001b).

The "social vision" proclaimed by Acceso identified three fundamental concepts: "equitable access" or the possibility to connect at a reasonable price and with basic training so that any person could benefit from Internet technologies and services; "meaningful use" or "effective use of ICT resources, both alone and in combination with other appropriate means of communication"; and "social appropriation" or the capacity to transform aspects of daily life through the use of technology to resolve specific problems (Gómez & Martínez, 2001b, pp. 6–7). In that way, aspects of infrastructure, participation practices, and concrete developmental indicators were linked together.

The purpose of the Acceso proposal was also to transcend the notion itself of the "digital divide" by understanding it "not [as] a cause but a manifestation of existing social, economic, and political divides at the local, national, and global levels" (Gómez et al., 2001, p. 111). Following that logic, approaches focused on resolving the "digital divide" were destined to fail in the sense that they isolated it from the factors that produced it. Hence, talking about "environments" that would "enable" developmental possibilities was preferred. This required better technology integration, active citizen participation in all aspects associated with technological development, and the integration of "an ethic of solidarity, reciprocity, and enthusiasm" into Internet use and production (Gómez & Martínez, 2001b, p. 9). These "environments," according to researchers affiliated with Acceso, permitted reaching the Internet's true potential: greater citizen participation, new ways of collaborating, and the inclusion of voices that had been relatively excluded from other ways of participating (such as social movements and indigenous peoples).

Given that the Internet was not seen as an intrinsically positive technology, the risks related to its use were also recognized, notably the possibility to increase inequalities instead of decrease them. Significantly, Acceso's analytical project included gender issues as a central axis for considering the design, use, and analysis of the relationship between technology and development.

One of the great expectations of those who participated in elaborating this array of ideas was to be able to have an impact on public policy formation "from the point of view of social transformation [that] [would] respond] most appropriately to the population's needs and to the agendas of civil society organizations" (Martínez, 2001b, p. 509). However, despite its close association with organizations and its concrete work experiences in the field, this hope was not achieved. Rather, public policies emerged in the era that prioritized the discourse of the "information societies" and "knowledge economies" that had found fertile ground in the region's governmental institutions. Though promising, the proposal was not the foundation of an emerging school of thought or Latin American theory about technology and development that would give analytical continuity to the body of concerns that crystalized in Latin America toward the end of the century. At the start of the new century, the theories followed a path similar to that of the technologies: they were overshadowed by the commercial concerns that dominated the conversation regarding the Internet.

CONCLUDING REMARKS

This chapter followed the transition of the Central American Internet from academia to the commercial world. To that end, the importance of privatization processes in redefining the Central American Internet field was discussed. The entrance of commercial Internet access providers expanded the array of options to connect in the region, but it substantially modified the work conditions and possibilities of interested actors to strengthen the network's use in sectors with lower market value. In that context, the project of integrating the region through computer networks began to lose meaning for actors such as universities and NGOs that needed to redefine their goals and methods. Thus, it could be concluded that the integrationist dream was sacrificed based on privatization and the opening of the telecommunications market.

One of the most distinctive features of the processes discussed in this chapter was the consolidation of an institutional Internet field. The network's development in the region during the second half of the 90s was marked by the emergence of nascent use cultures (reflected by the appearance of practices such as "navigating" the Web and new ways of referring to these practices), means of organizing access (telecentres, cybercafés), institutional actors (regulatory agencies, commercial access providers), visions regarding the network's political significance for encouraging or limiting development, early indications of public policy, metaphors to understand the rise of technologies like the Internet and the mobile phone ("information societies" and "knowledge economies"), theories and bodies of knowledge to give the network direction… In short, the Internet acquired institutional strength and life.

A central argument in this book is that the formation of knowledge about the Internet cannot be separated from developments in the field of telecommunications. In other words, the institutional transformations and the theories that explain them are the result of the same historical conditions. To demonstrate this point, this chapter provided an archaeology of one of the first Latin American theories regarding the relationship between technology and development (ICT4D). This theory proposed going beyond access and connectivity and, instead, to create "environments" with the appropriate conditions to foster "meaningful use" of the Internet to the benefit of the Central American population. The common trajectory of this theory and of the process of integration through computer networks indicate a central theme present throughout this book: institutional concerns have been predominantly focused on the network's commercial value more than on encouraging meaningful uses for different segments of the region's population. The implications of this process are analyzed in the following and final chapter.

Note

1. At Fundación Acceso, the team interested in telecommunications matters included Nora Galeano, Juliana Martínez, and María Sáenz. After exploring Latin American initiatives, Ricardo Gómez, IDRC researcher, discovered the Foundation's work and established a link with the initiative. Thus, a productive academic collaboration emerged with the Fundación Acceso researchers that involved a short business trip to Costa Rica. Kemly Camacho joined the research team toward the start of the new century.

References

Abbate, J. (2000). *Inventing the internet*. Cambridge, MA: MIT Press.

Aguilar, C., & Ballestero, A. (1993). *Evaluación y alternativas de rediseño de la red pública de datos RACSAPAC*. Tesis presentada como requisito para optar por el grado de Magister Scientiae en Computación, Instituto Tecnológico de Costa Rica.

Ansorena, C. (2008). *Competencia y regulación en las telecomunicaciones: El caso de Nicaragua*. Mexico: CEPAL.

Argumedo, P. (2007). *Competencia y regulación en las telecomunicaciones: El caso de El Salvador*. Mexico: CEPAL.

Belt, J. A. B. (1999). Telecommunications reform to promote efficiency and private sector participation: The cases of El Salvador and Guatemala. *Economists Working Paper Series, 10*, 1–10.

Bonilla, M., & Cliche, G. (Eds.). (2001). *Internet y sociedad en América Latina y el Caribe: Investigaciones para sustentar el diálogo.* Quito, Ecuador: FLACSO.

Bull, B. (2005). *Aid, power and privatization: The politics of telecommunication reform in Central America.* Cheltenham, UK: Edward Elgar.

Bush, R., & Meyer, D. (1997). InterRed Panama. *Psg.com*, from https://archive.psg.com/970810.intered/sld001.htm

CEPAL. (2003). *Los caminos hacia una sociedad de la información en América Latina y el Caribe* (LC/G.2195). México: CEPAL.

de Téramond, G. (1996a). Consolidación del backbone Internet en Centroamérica. Paper presented at Primer Foro Regional: La universidad centroamericana hacia el tercer milenio, San José, Costa Rica.

de Téramond, G. (1996b). Internet en Costa Rica. *La Nación*, from https://www.nacion.com/opinion/internet-en-costa-rica/FPIAKZ7MARCPJJUP3ZFWNQ4C7Q/story/

de Téramond, G., & Brenes, A. (1995). *Establecimiento y consolidación de los proyectos Internet en Centro América y el Caribe bajo el marco del proyecto RedHUCyT. Etapa II: Hacia el establecimiento de un backbone regional.* San José, Costa Rica: CRNet.

de Téramond, G., Brenes, A., Espinoza, D., & Bonilla, R. (1997). *Establecimiento y consolidación de los proyectos Internet en Centro América y el Caribe bajo el marco del proyecto RedHUCyT. Etapa IV: Extensión de los proyectos e interconectividad regional.* San José, Costa Rica: CRNet.

Fervoy, P., Martínez, J., & Sáenz, M. (2000). Promoting equitable access, meaningful use and appropriation of the internet: Recommendations for ECOSOC. In S. Khan & D. Gold (Eds.), *Information and communications technology and development in the new millennium: Dialogues at the United Nations* (pp. 91–96). New York: United Nations.

Fumero, G. (2013). *Telecomunicaciones en Costa Rica: 140 años de historia en defensa de un servicio público.* San José, Costa Rica: EUNED.

Furlán, L. (2007, January 9). *Guatemala: Una pequeña historia de Internet.* Retrieved from https://interred.wordpress.com/2007/01/09/una-pequena-historia-de-internet-en-guatemala/

Gómez, R., & Barnola, L. G. (Eds.). (2000). *Tacking stock: Lessons and experiences of the Pan@Americas networking program.* Picton, ON: IDRC.

Gómez, R., & Martínez, J. (2001a). *Central America: Towards a social use of the internet.* Ottawa, Canada: IDRC.

Gómez, R., & Martínez, J. (2001b). *The internet… Why? And what for?* Ottawa, Canada: IDRC.

Gómez, R., Martínez, J., & Reilly, K. (2001). Paths beyond connectivity: Experience from Latin America and the Caribbean. *Cooperation South, 1*, 110–122.

González, R. (2007). *Competencia y regulación en las telecomunicaciones: El caso de Panamá*. Mexico: CEPAL.

Haglund, L. (2006). Hard pressed to invest: The political economy of public sector reform in Costa Rica. *Revista Centroamericana de Ciencias Sociales, 3*(1), 5–46.

Hahn, S. (1999). Case studies on developments of the internet in Latin America: Unexpected results. *Bulletin of the American Society for Information Science, 25*(5), 15–17.

Hahn, S. (2000, July 18–21). *Case studies on development of the Internet in Latin America and the Caribbean*. Paper presented at the Internet Society (INET) conference, Yokohama, Japan.

Hoffmann, B. (2004). *The politics of the internet in third world development*. London: Routledge.

Hoffmann, B. (2008). Why reform fails: The "politics of policies" in Costa Rican telecommunications liberalization. *European Review of Latin American and Caribbean Studies, 84*, 3–19.

Martínez, J. (2001a). Central America: National environments for internet access. In *Internet and society: Reflecting on public policies* (Draft no. 7). San José, Costa Rica: Fundación Acceso.

Martínez, J. (2001b). Internet y políticas públicas socialmente relevantes: ¿Por qué, cómo y en qué incidir? In M. Bonilla & G. Cliche (Eds.), *Internet y sociedad en América Latina y el Caribe: Investigaciones para sustentar el diálogo* (pp. 509–541). Quito, Ecuador: FLACSO.

Monge, R. (2000). La economía política de un intento fallido de reforma en telecomunicaciones. In R. Jiménez (Ed.), *Los retos políticos de la reforma económica en Costa Rica* (pp. 273–318). San José, Costa Rica: Academia de Centroamérica.

Monge, R., & Hewitt, J. (2004). *Tecnologías de la información y las comunicaciones (TIC) y el futuro de Costa Rica. El desafío de la exclusión*. San José, Costa Rica: CAATEC.

Pasch, G. (1996). *América Central: Proveedores comerciales de servicios electrónicos*. 9 de Agosto de 1996, from https://web.archive.org/web/19961018052642/http://fiat.gslis.utexas.edu/~gpasch/sproveen.html

Pasch, G., & Valdés, C. (1997, February 3–5). *The dawn of the internet era in Guatemala*. Paper presented at the international federation of information processing, Florianopolis, Brazil.

Proenza, F. J., Bastidas-Buch, R., & Montero, G. (2001). *Telecentros para el desarrollo socioeconómico y rural en América Latina y el Caribe*. Washington, DC: University Institute of Technology.

Raventós, P. (1998). *Telecommunications in Central America* (Development discussion paper no. 648). Harvard University.

Raventós, P. (2001). Deregulating telecommunications in Central America. In F. Larraín (Ed.), *Economic development in Central America* (Vol. II: Structural reform) (pp. 107–156). Cambridge, MA: Harvard University Press.

RDS. (s.f.). Antecedentes. *Red de Desarrollo Sostenible Honduras*, from https://rds.hn/perfil/

Rivera, E. (2007). *Modelos de privatización y desarrollo de la competencia en las telecomunicaciones de Centroamérica y México*. Mexico: CEPAL.

Rodríguez, M. A. (2001). Incorporando a Costa Rica en el Mundo Digital. In CAATEC (Ed.), *Costa Rica en el mundo digital: Retos y oportunidades*. San José, Costa Rica: CAATEC.

Sáenz, M., & Galeano, N. (1996). *Estado de las telecomunicaciones en Centroamérica: Impresiones sobre la situación actual*. San José, Costa Rica: Fundación Acceso.

SDNP. (1997). *SDNP country reports*. United Nations Sustainable Development Networking Programme.

Siles, I. (2008). *Por un sueño en.red.ado. Una historia de Internet en Costa Rica (1990–2005)*. San José, Costa Rica: Editorial de la Universidad de Costa Rica.

Siles, I. (2012). Establishing the internet in Costa Rica: Co-optation and the closure of technological controversies. *The Information Society, 28*(1), 13–23.

Siles, I. (2017). 25 years of the internet in Central America: An interview with Guy de Téramond. *Internet Histories, 1*(4), 349–358.

Siles, I., Espinoza, J., & Méndez, A. (2016). ¿El Silicon Valley latinoamericano?: La producción de tecnología de comunicación en Costa Rica. *Anuario de Estudios Centroamericanos, 42*, 411–431.

Tábora, M. R. (2007). *Competencia y regulación en las telecomunicaciones: El caso de Honduras*. Mexico: CEPAL.

Trejo Delarbre, R. (1996). *La nueva alfombra mágica: Usos y mitos de Internet, la red de redes*. Madrid, Spain: Fundesco.

Urízar, C. (2007). *Competencia y regulación en las telecomunicaciones: El caso de Guatemala*. Mexico: CEPAL.

CHAPTER 7

The Inconclusive Project of Technological Integration

Abstract This chapter discusses the implications of the evidence presented in the book. First, it emphasizes the significance of transnational perspectives to examine Internet histories. Specifically, it shows how a transnational approach invites a reconsideration of established traditions in the historical analysis of the Internet. Second, it argues that processes of technological integration examined in the book remain an unfinished project. The book concludes by showing how the history of the Internet in Central America helps to think about present issues, such as the uneven access to computer networks and its implications of understanding the development of the region.

Keywords Central America • Computer networks • Development • Integration • Internet • Transnational history

The previous chapters gave an account of the processes that brought forth regional Internet connection in Central America in the first half of the 1990s. A transnational focus was used to explain dynamics at a local level. In that way, the discussion was about how the region's countries began networking projects at the end of one of the most difficult decades in its history. By the mid-90s, Central American Internet began to make room for local and national networks, and the process of integrating the region through networking technologies began to fade.

© The Author(s) 2020
I. Siles, *A Transnational History of the Internet in Central America, 1985–2000*, Palgrave Macmillan Transnational History Series, https://doi.org/10.1007/978-3-030-48947-2_7

135

In light of the evidence presented in the previous chapters, this concluding chapter revisits the principal arguments developed in this book. Two particular topics will be addressed: (a) the relevance of a transnational approach for analyzing Internet histories, and (b) the network's political value as a means to integration and development.

Toward Transnational Internet Histories

The early development of computer networks in the global south has principally been explained as the result of these networks' low cost as well as their political and technological characteristics (Murphy, 2004). Thus, Abbate (2000) stated that, outside of the U.S., "grassroots user-supported networks with lower political profiles, such as BITNET and UUCP, spread faster than the Internet" (p. 209). Furthermore, this book demonstrated that, in addition to these factors, establishing these networks was influenced by the configuration of transnational processes among Central American countries. These processes can be analyzed based on three transnationalist dimensions presented in this book's introduction: (a) flows of people, knowledge, and technologies; (b) the role of specific actors (such as international organizations and "network entrepreneurs"); and (c) the consequences associated with the study of these transnational processes in configuring Internet projects at the local level.

Firstly, establishing computer networks in Central America, at least initially, was the result of flows of people, knowledge, and technologies. These flows are not only a current feature of the Internet but a condition of its historical development. The implementation of these networking projects required navigating the particular technological, political, and institutional context that characterized Central America's reconstruction at the start of the 1990s. Each phase of the process, such as getting to know early computer networks, negotiating with state telecommunications monopolies, forming academic networks to share costs, establishing satellite links, searching for new users, and receiving training in the maintenance and growth of these networks, was marked by transnational flows. To understand these processes, it was essential to consider how people (engineers and collaborators, representatives of international entities and transnational enterprises, public officials), knowledge (protocol management, website configuration), and technologies (UUCP, BITNET, the Internet, routers, modems) were mobilized across national borders.

In that sense, computer networks can be viewed as a result (and not only a cause) of transnational exchange networks. In other words, the Internet not only permitted maintaining academic relationships between people in different contexts, as its promoters had hoped, but it was also a result itself of these same types of collaborations. For the participants of these networking projects, in addition to an array of applications and protocols, the Internet was the way to develop these exchanges with regional colleagues.[1] Hence, computer networks provided the possibility to expand this transnational network of scientific collaboration over time and space.[2] This perspective also allows for an understanding of the differences between the region's institutional projects. The Internet, as a technology, was the materialization of different Latin American political projects. In practice, establishing permanent network access was viewed by local promoters as a greater good that merited reconciling institutional interests that came into conflict with each other throughout the process.

Various specific conditions facilitated these transnational flows. To start, the signing of peace agreements created favorable political conditions for regional integration projects. Additionally, Central America's size and geography made it possible to consider solutions in regional terms. In other words, at least in what telecommunications refers to, the "geomorphic particularities" of the isthmus did in fact contribute to regional and integrationist projects (Pérez Brignoli, 2017, p. 4). Several of the proposed Internet-development projects—including direct connection between Costa Rica, Nicaragua, and Panama—would have been more challenging to develop in other places, even in Latin America. At the same time, certain regional infrastructure (such as the Regional Technical Telecommunications Commission microwave network or the X.25 network itself, over which the former operated) laid the foundation that made it possible to expand UUCP and develop projects such as the "Central American Internet backbone."

Secondly, this transnational perspective made the major role of certain actors clear. Numerous international organizations were key in that sense. The work of the Superior Central American University Council encouraged adopting UUCP, and IBM's regional presence facilitated expanding BITNET. It is precisely because of this action capacity that organisms such as the Organization of American States and the UN Development Programme could both obtain and distribute resources in the region. The transnational nature of these organizations made it possible to negotiate politically with local governments, telecommunications enterprises, and academic entities to acquire and install equipment in the region. This was recognized repeatedly by the participants of every networking project. Abel Brenes, technical director of CRNet and one of the Costa Rican engineers that traveled throughout Central America, maintains, "Everything worked because the OAS was there [...] to integrate and to smooth over [working relationships]" (interview with the author, July 4, 2017). This is not a trivial factor in the political and economic context of 1990s Central America.

International organizations found ideal allies in the "network entrepreneurs" like de Téramond, Hahn, Hope, and Puliatti. These people played a key role in establishing transnational "contact zones" (Sandoval García, 2009), such as conferences, meetings, seminars, and workshops, where interested actors found potential collaborators from other countries, designed solutions, and reached agreements. They also traveled around the entire Central American region to meet with counterparts, quell criticism, establish Internet access, and train local counterparts.

Finally, the focus adopted in this study helped to better comprehend the consequences of transnational processes in terms of how the Internet was implemented at the local level. In each country, transnational exchanges informed local projects in particular ways. From that point of view, Central American Internet history cannot be only told from a national perspective. Certainly, the local and national history of each of these countries are possible (and have been written) (Pasch & Valdés, 1997; Siles, 2008, 2012). These histories have recognized the international support and relationships that were needed to establish Internet in each country. However, what a national angle fails to make visible is the constitutive nature of those transnational exchanges. Through those flows and exchanges, a regional Internet became possible. In that sense, it makes just as much sense to speak of a Central American Internet in the first half of the 1990s as it would to refer to national Internet(s) in Central America.

For obvious reasons, the network as a figure has occupied a prominent position in discussions of the Internet's history. Nevertheless, more attention has been given to technological or computational networks than to the networks of individuals who collaborated in different countries and organizations to establish the Internet. This transnational investigation demonstrated that the Internet is as much a collaborative network as it is a computer network defined by the use of certain protocols. As discussed in previous chapters, in Latin America it became common to speak of "computer networks" and "human networks" to make visible the need to differentiate both (Silvio, 1992).

Enabled by organizations with a regional presence, transnational networks of exchange and collaboration were established to implement and circulate computer networks in Central America. This allowed for the consolidation of projects strengthened by the Superior Central American University Council, the Organizations of American States, and the UN Development Programme: networks of regional counterparts were needed to give support to computer networks, and keep them alive. Once implemented, certainly, these computer networks also shaped the nature of regional collaborators' exchange networks. The history of collaborative "human" networks that made the Internet what it has become, is quite less well-known than the history of its technologies and protocols, particularly outside of the U.S. One of this book's purposes was to aid in reversing this tendency.

Of "Regional Arteries" and "Backbones" for Development

Analyzing the transnational flows that traversed Central American countries from 1985 to 2000 made it possible to identify computer networks' political significance as a means to regional integration. This aspect has scarcely been taken into account in academic literature about Central American history and regional integration processes. Viewed in this way, Central American networking initiatives indicated an exit from the region's "lost decade" through a particular means of technological integration. Though clearly linked to institutional projects of economic and political development, this type of integration must be discussed on its own terms.

As was clear from the transition between a "regional artery" in the 60s and an Internet "backbone" in the 90s, it is necessary to comprehend the way in which integration and technology have been mutually defined throughout history: visions of integration (of an "introspective" or "open" type, for example) have permeated technological development projects, and, at the same time, technologies (such as a microwave network or computer networks like the Internet) have given a unique and tangible expression to integrationist projects in Central America.

Examined in previous chapters, networking projects implemented and promoted a particular vision of integration. This approach centered on forming sociotechnical networks that would allow for the circulation of people, knowledge, and technologies among the region's countries. In other words, rather than eradicating national borders, these projects sought to cross them. In that way, the hope was to establish not only the Internet but also human networks, knowledge(s), and infrastructures that would be strong enough to give these computer networks viability, magnitude, and possibilities for growth in the mid- and long term. These efforts came from projects based out of universities (and several civil sector organizations) that were less interested in the institutional dimension of the integration project but that saw in it "the gate to balanced and self-sustained regional development" (Cerdas, 2005, p. 205).

In that sense, this book argued that results of this type of technological integration can also be evaluated from a developmental perspective. During the first half of the 90s, the circulation of people, knowledge, and technologies facilitated institutionalizing a means of transnational work that had significant repercussions for Central American academia. In that way, local teams were established with the capacity to give direction to the network's development based on their academic work. In contrast to other parts of the world, this gave the Central American region the autonomy to decide the paths that computer networks like the Internet would follow in the coming years.

At the same time, the institutional or ecological perspective adopted in this investigation also made it possible to identify the integrationist project's limitations for contributing to regional development. As proposed in the previous chapter, it is thus essential to situate the Internet's implementation and growth in the context of privatization dynamics. The fragmentation of the transnational push for the Internet is a direct effect of this privatization insomuch as it limited the integrationist agenda. Commercial exploitation of the network detracted from the search for shared solutions to common problems. In that way, the joint project of constructing infrastructure that would make it possible for the region to exchange communication without having to leave its territory was abandoned. Almost 25 years after its original proposal, the project of implementing a Central American backbone remains inconclusive.

Another void that this investigation has made clear relates to the deficiencies in consolidating public policies in the region. At the turn of the century, only Costa Rica had a policy come into being that, despite its echoes of commercial discourse regarding the Internet's value at the expense of other dimensions, sought to encourage its use as a means to foster development. Public policy can make a significant difference in the creation of appropriate conditions for achieving "meaningful [technological] use"—in this sense given by the pioneering theory that emerged in Latin America in the late 90s (described in Chap. 6). At the same time, the previous chapters suggest that a deeper or more systemic institutional reconfiguration is needed. The search for concrete solutions to strengthen development (such as public policy) seems limited when the institutional field, as a whole, is oriented toward other purposes.

In more ways than one, the current state of the Internet in Central America is a reflection and result of the historic conditions discussed in this book. According to Internet World Stats (2017), on average, only 50.6% of the region's population had Internet access in 2017. Additionally, access is distributed unequally. Only Costa Rica and Panama have numbers higher than this average (86.4% in Costa Rica and 69.1% in Panama), while in El Salvador, Internet access is practically equal to the regional average (50.3%) and in Guatemala (34.5%), Honduras (32.5%) and Nicaragua (30.6%), it is significantly inferior. The figures for network access in Costa Rica reflect the significance that mobile Internet acquired in the country. This country has the greatest mobile Internet access of Latin America and the Caribbean (CEPAL, 2016). The relatively low cost of cellular telephone services would be a key factor in this development.

In all of the region's countries, higher-income and urban areas have experienced higher growth in Internet use since 2010 than lower-income and rural areas. Only in Costa Rica has the access divide between rural and urban zones not grown in the present decade. Although Internet access costs have decreased (when comparing monthly GDP per capita), in the region's countries, most notably in Nicaragua, costs are still relatively high (CEPAL, 2016). Beyond the numbers, the recent advance to encourage "meaningful use" of the network has been modest. Over the 2010 decade, several reforms were implemented to encourage this type of Internet use in countries like Costa Rica (National Telecommunications Development Plan, 2015–2021), Honduras (Honduran Digital Agenda, 2014–2018) and Panama (Digital Agenda, 2014–2019). All of these plans seek to promote "meaningful uses" of the network as one of their principal goals, but all have encountered political difficulties to achieving this objective.

Finally, the evidence in this book shows that, similar to the technological integration project of the 90s, academic knowledge—in the sense of theories and conceptual models—about the Internet as a means to strengthen (or obstruct) development, stopped taking into account regional terms. Few transnational exchanges exist that make it possible to understand the region's situation beyond a comparison of access statistics. Despite its impetus at the start of the new century, the regional perspective provided by one of the first Latin American (and global) theories about the relationship between technology and development began to fade. The academic conversation about the Internet's value seems to be heavily dictated by those from other latitudes (Gómez-Cruz & Siles, 2020; Siles, Espinoza, & Méndez, 2019; Waisbord, 2014). In that context, this book hopes to have contributed, through a historical seed, to a better understanding of why the Internet matters in Latin America, in general, and Central America, in particular.

Notes

1. Technologies that circulated in the region before Internet, such as magnetic discs and cassettes, could also be analyzed from this perspective.
2. Latour (1991) holds a similar argument in evaluating the effect of socio-technical networks on maintaining relationships over time and space.

References

Abbate, J. (2000). *Inventing the internet.* Cambridge, MA: MIT Press.

CEPAL. (2016). *Estado de la banda ancha en América Latina y el Caribe 2016.* Santiago, Chile: Naciones Unidas.

Cerdas, R. (2005). *Las instituciones de integración en Centroamérica: De la retórica a la descomposición.* San José, Costa Rica: EUNED.

Gómez-Cruz, E., & Siles, I. (2020). Digital cultures. In W. Raussert, G. L. Anatol, S. Thies, S. Corona & J. C. Lozano (Eds.), *The Routledge handbook to the cultures and media of the Americas* (pp. 319–329). London: Routledge.

Internet World Stats. (2017). Internet usage and population in Central America. *Internet World Stats.* Retrieved from https://www.internetworldstats.com/stats12.htm

Latour, B. (1991). Technology is society made durable. In J. Law (Ed.), *A sociology of monsters: Essays on power, technology and domination* (pp. 103–131). London: Routledge.

Murphy, B. M. (2004). Propagating alternative journalism through social movement cyberspace: The appropriation of computer networks for alternative media development. In R. Eglash, J. L. Croissant, G. Di Chiro, & R. Fouché (Eds.), *Appropriating technology: Vernacular science and social power* (pp. 163–180). Minneapolis, MN: University of Minnesota Press.

Pasch, G., & Valdés, C. (1997, February 3–5). *The dawn of the internet era in Guatemala*. Paper presented at the international federation of information processing, Florianopolis, Brazil.

Pérez Brignoli, H. (2017). *El laberinto centroamericano: Los hilos de la historia*. San José, Costa Rica: CIHAC.

Sandoval García, C. (2009). Zonas de contacto entre las ciencias sociales. In M. Baltodano, E. Cook, & R. Mooney (Eds.), *Género y religión: Sospechas y aportes para la reflexión* (pp. 177–191). San José, Costa Rica: Editorial SEBILA.

Siles, I. (2008). *Por un sueño en.red.ado. Una historia de Internet en Costa Rica (1990–2005)*. San José, Costa Rica: Editorial de la Universidad de Costa Rica.

Siles, I. (2012). Establishing the internet in Costa Rica: Co-optation and the closure of technological controversies. *The Information Society, 28*(1), 13–23.

Siles, I., Espinoza, J., & Méndez, A. (2019). La investigación sobre tecnología de comunicación en América Latina: Un análisis crítico de la literatura (2005–2015). *Palabra Clave, 22*(1), 12–40.

Silvio, J. (Ed.). (1992). *Calidad, tecnología y globalización en la educación superior latinoamericana*. Caracas, Venezuela: CRESALC-UNESCO.

Waisbord, S. (2014). United and fragmented: Communication and media studies in Latin America. *Journal of Latin American Communication Research, 4*(1), 55–77.

Index[1]

[1] Note: Page numbers followed by 'n' refer to notes.

© The Author(s) 2020   145
I. Siles, *A Transnational History of the Internet in Central America,
1985–2000*, Palgrave Macmillan Transnational History Series,
https://doi.org/10.1007/978-3-030-48947-2